Seed Biology and Yield of Grain Crops, 2nd Edition

This book is dedicated to four people who, in one way or another, made it possible. My mother Florence Egli, my major professor Dr J.W. Pendleton, University of Illinois, and two long-term colleagues at the University of Kentucky – the late W.G. Duncan and J.E. Leggett.

Seed Biology and Yield of Grain Crops, 2nd Edition

Dennis B. Egli
University of Kentucky
USA

CABI is a trading name of CAB International

CABI	CABI
Nosworthy Way	745 Atlantic Avenue
Wallingford	8th Floor
Oxfordshire OX10 8DE	Boston, MA 02111
UK	USA
Tel: +44 (0)1491 832111	Tel: +1 (617)682-9015
Fax: +44 (0)1491 833508	E-mail: cabi-nao@cabi.org
E-mail: info@cabi.org	
Website: www.cabi.org	

A catalogue record for this book is available from the British Library, London, UK.

Library of Congress Cataloging-in-Publication Data

Names: Egli, Dennis B.
Title: Seed biology and yield of grain crops / Dennis B. Egli.
Description: 2nd edition. | Wallingford, Oxfordshire : CAB International,
  [2017] | Includes bibliographical references and index.
Identifiers: LCCN 2016035557| ISBN 9781780647708 (hbk : alk. paper) | ISBN
  9781780647722 (epub)
Subjects: LCSH: Grain--Seeds--Physiology. | Grain--Yields.
Classification: LCC SB189.4 .E45 2017 | DDC 633.1/04--dc23 LC record available at
https://lccn.loc.gov/2016035557

ISBN-13: 978 1 78064 770 8

Commissioning editor: Rachael Russell
Editorial assistant: Emma McCann
Production editor: Shankari Wilford

Typeset by SPi, Pondicherry, India
Printed and bound in the UK by CPI Group (UK) Ltd, Croydon, CR0 4YY, UK

Contents

Preface

The world's food supply depends on crops harvested for their seeds. Roughly half of the calories available from plant sources in recent years came from just four crops harvested for their seeds – maize, rice, wheat and soybean. Seeds are harvested because they are rich in carbohydrate, protein and oil stored in the seed as reserves for germination and the beginning of the next generation. Dry seeds are easy to transport and store; characteristics that contribute to their usefulness and popularity.

The unique carbohydrates, proteins and oils in the seed result from a complex series of biochemical processes, starting with the capture of light energy and the fixation of carbon in the leaf and ending with the synthesis of storage compounds in the seed. The mother plant produces the raw materials, primarily sucrose and various amino acids that are used by the seed to synthesize the complex molecules we use as food or feed. Understanding the production of yield by a crop community requires consideration of both the assimilatory and the synthesis processes.

Crop physiologists historically focused on the assimilatory processes. Investigations of dry matter accumulation by plants and plant communities and photosynthesis and other primary assimilatory processes were considered important because these processes are fundamental to the production of yield. However, the production of dry matter by a crop community is only part of the story in a grain crop where the economic yield is the seed. Utilization by the seed of raw materials translocated from the source is an equally important part of the yield production process. That is what this book is about.

My objectives in this book are, first, to gain an understanding of the growth and development of seeds, the processes involved, the regulation of these processes and the effect of plant and environmental factors. The second objective is to use this knowledge of seed growth and development to define the role of the seed in the yield production process.

What will we gain from such considerations? By approaching the production of yield from the viewpoint of the accumulation of dry matter by the seed (the sink), we will be able to integrate the source and the sink, assimilatory and synthesis processes, into a unified description or model of yield production. This model will be better than one that considers only the assimilatory processes in the source and relegates sink activity to a black box. A unified model including the seed will help us understand many important questions in yield physiology, including the determination of seed number, the relationship between seed size and yield, partitioning and source-sink relations. We cannot hope to answer all questions about the regulation of yield in a single book, but a thorough consideration of the seed sink will contribute to that goal.

Acknowledgements to First Edition

Preparation of a book purporting to cover such a broad topic as seed growth and development and the involvement of the seed in the production of yield of grain crops relies primarily on the published and unpublished research of others. Synthesis of information from the literature into principles and concepts occurs over a period of many years and it is frequently difficult to remember where certain ideas or concepts originated – did they come from a long-forgotten paper or talk at a conference, or are they mine? I find it sometimes difficult to answer this question and therefore can only issue a blanket acknowledgment and express my thanks to those whose ideas I inadvertently expropriated and used and could not cite specifically. This book also draws heavily on my own research over the past 20 years and that research was made possible by the able assistance of many graduate students, technicians, visiting professors and colleagues. My heartfelt thanks go to this group.

A number of individuals made specific contributions to this book and I greatly appreciate their efforts. Dr Steve Crafts-Brandner, USDA-ARS, Western Cotton Research Laboratory, my valued colleague for many years at the University of Kentucky, graciously made it possible for me to spend a six-month sabbatical with him in Phoenix, Arizona, where I had the solitude to concentrate on writing. The final version of this book was completed during my stay in Phoenix. My thanks also goes to the other scientists and staff at the Western Cotton Research Laboratory for making my stay there most enjoyable. Mr Bill Bruening ably managed my lab in Lexington during my absence and prepared most of the figures in this book. Brenda Wilson prepared most of the tables. I thank Lynn Forlow Jech, Western Cotton Research Laboratory, for preparing Fig. 4.6. Individuals that reviewed chapters include Steve Crafts-Brandner, Jim Heitholt, Miller MacDonald, Dennis TeKrony and Ian Wardlaw. Their comments were insightful and very helpful and I appreciate their efforts, but the opinions expressed in this book are solely the responsibility of the author.

Dennis B. Egli
September 1997
Lexington, Kentucky

Acknowledgements to Second Edition

This second edition of *Seed Biology and the Yield of Grain Crops* incorporates the substantial literature on seed growth and development and its role in determining yield that accumulated in the nearly 20 years since publication of the first edition in 1998. As before, it is not possible to cite all of the papers contributing to the ideas and concepts developed in this book and, as usual, it is not always clear where new ideas and concepts originated, so I apologize to those whose work was not cited and to those that may not have received the credit they deserve.

My research over the last 20 years would not have been possible without the able assistance of two research analysts, Bill Bruening and Marcy Rucker; I express my sincere appreciation to them for keeping my lab operating at peak efficiency at all times. I also want to thank the innumerable graduate students and my colleagues in the Plant and Soil Sciences Department for their contributions to my research programme. A special thank you to H.M. Davies, Professor in the Plant and Soil Sciences Department at the University of Kentucky, for our many enlightening discussions on the involvement of biotechnology in yield improvement and for his review of an early version of Chapter 6. He willingly contributed his photographic expertise to convert the photos in this book to a digital format. Finally, I want to express my appreciation to the Plant and Soil Science Department and the University of Kentucky for my phased retirement appointment that gave me the time to complete this book.

Dennis B. Egli
June 2016
Lexington, Kenucky

Introduction

1

Seeds as a Food Source

Humans have always relied on the green plant to produce the calories needed for their sustenance, either directly or indirectly after conversion by animals, and as a source of fuel and fibre. As a result of this reliance on green plants, the sun was essentially the only source of energy until the exploitation of fossil forms of solar energy ushered in the industrial revolution. Agricultural production systems became increasingly dependent upon these fossil forms of energy (coal, petroleum), but solar energy, diffuse but reliable, continued to be the primary source of our food supply (Hall and Kitgaard, 2012, p. 4). The green plant driven by solar energy will, for the foreseeable future, continue to feed humankind.

The plants utilized by humans are consumed in many different ways; for some, fresh fruits are harvested, in other cases stems, leaves, roots or tubers represent the economic yield. The entire above-ground plant is harvested in some vegetable or forage crops whereas immature fruits or seeds represent the economic yield of other vegetable crops. But the crop plants making the largest contribution, by far, to the world's food supply, are those harvested at maturity for their seed.

Seeds are important and useful because they are nutrient-dense packages of carbohydrates, protein and oil that are relatively easy to harvest, store and transport. Once the seed is dried, it can be stored indefinitely if it is kept dry and free of insects and other pests. Storage of seed is cheaper and the shelf-life is infinitely longer than plant parts that are consumed fresh. Its ease of transport provided the foundation of the global grain trade that has helped equalize worldwide supply and demand since the development of ocean-going ships (originally moved by solar energy in the form of wind). Seeds are an important source of animal feed to produce meat, eggs, milk and other animal products.

The seed is also the biological unit used to reproduce most crops; there would be little food production without adequate supplies of viable, vigorous planting seed. The slogan of the American Seed Trade Association – 'First the Seed' – makes it clear that our existence depends on seeds that can germinate to produce the next crop. Thus, seed has a dual function of being consumed as food or feed and providing the means to reproduce the crop. These attributes have made the seed the foundation of agriculture since ancient times.

Many plant species have been used as sources of food, feed or fibre. Harlan (1992) compiled a 'short list' of cultivated plants that contained 352 species from 55 families. Vaughan and Geissler (1997) listed approximately 300 plant species used for food. The database of agricultural statistics (FAOSTAT) of the Food and Agriculture Organization (FAO) of the United Nations lists some 130 species in their crops category including grains, vegetables, fruits, nuts, fibre crops, spices and stimulants (coffee, tea and tobacco), but seeds are harvested from only about 35 species (FAOSTAT, 2014) and only 22 of these species are produced in substantial amounts (Table 1.1).

These 22 species represent only a few families, with 18 of them from the *Poaceae* (grasses) (nine) and the *Fabaceae* (legumes) (nine). Three of the species (maize, rice and wheat) dominate the world grain (seed) production, accounting for 76% of the 2011–2014 average production of the species in Table 1.1. If soybean, the fourth major crop, is included, the total increases to 84%. These crops account for roughly half of the calories available per capita for consumption from plant sources in 2009–2011. This proportion would increase if the seeds fed to livestock were included. It is clear that humans are fed by a very small sample of the plant species that could be used to produce food. Relying on so few crop species would seem to make our food supply vulnerable to insect or disease epidemics, but the use of multiple varieties of each crop reduces the chances of widespread crop failure (Denison, 2012, p. 3) as does the worldwide distribution of each crop. The importance of maize, rice and wheat is not a recent phenomena; Heiser (1973) pointed out that most important early civilizations were based on seeds of these crops. Truly, crops harvested for their mature seeds have served us well.

There is continuing interest in increasing the number of plant species providing our food supply. Examples of new crop species under consideration include grain amaranth (*Amaranthus* spp.) (Gelinas and Seguin, 2008), chia (*Salvia hispanica* L.) (Jamboonsri *et al.*, 2012), quinoa (*Chenopodium quinoa*), hemp seed (*Cannabis sativa* L. (Pszczola, 2012), vernonia (*Vernonia galamensis*) (Shimelis *et al.*, 2008), and potato bean (*Apios americana* sp.), a legume that produces edible tubers (Belamkar *et al.*, 2015). Attempts are also being made to develop perennial grains from conventional annual crops and exotic species. Perennial grain crops are expected to conserve soil resources by providing continuous ground cover and perhaps produce higher yield as a result of a longer life cycle (Glover *et al.*, 2010).

New crops are often touted on the basis of their superior nutritive characteristics and/or their ability to be productive on infertile or droughty soils. If these new species are, in fact, 'super crops', why were they not selected in the long domestication processes that produced the few crops that feed the world? Are the species currently used those best suited for domestication (Sinclair and Sinclair, 2010, pp. 15–23), or were they domesticated first and then simply maintained by humans' unwillingness to start over (Warren, 2015, pp. 164–167)? The relatively poor track record of new crop development schemes in recent times suggests that there may

Table 1.1. World production and seed characteristics of crops where the mature seed is harvested for food or feed.

Crop		World production[1] (1000 t)	Harvested unit	Seed composition[2]		
				Carbohydrate (g kg^{-1})	Oil (g kg^{-1})	Protein (g kg^{-1})
Poaceae						
Maize	*Zea mays* L.	950,394	Caryopsis	800	50	100
Rice	*Oryza sativa* L.	733,424	Caryopsis	880	20	80
Wheat	*Triticum spp.*[3]	700,828	Caryopsis	750	20	120
Barley	*Hordeum vulgare* L.	138,252	Caryopsis[4]	760	30	120
Sorghum	*Sorghum bicolor* (L.) Moench	58,647	Caryopsis	820	40	120
Millet[5]	*Panicum miliaceum* L.	26,528	Caryopsis	690	50	110
Oat	*Avena sativa* L.	22,639	Caryopsis[4]	660	80	130
Rye	*Secale cereale* L.	14,906	Caryopsis	760	20	120
Triticale	*X Triticosecale Wittm ex A. Camus*	14,653	Caryopsis	594	18	131
Fabaceae						
Soybean	*Glycine max* (L.) Merrill	272,426	Non-endospermic seed	260	170	370
Groundnut[6]	*Arachis hypogaea* L.	41,366	Non-endospermic seed	120	480	310
Bean[7]	*Phaseolus vulgaris* L.	23,898	Non-endospermic seed	620	20	240
Chickpea	*Cicer arietinum* L.	12,735	Non-endospermic seed	680	50	230
Pea, dry[8]	*Pisum sativum* L.	11,013	Non-endospermic seed	520	60	250
Cowpea	*Vigna unguiculata* (L.) Walp.	6,661	Non-endospermic seed	570	10	250
Lentil	*Lens culinaris* Medikus	4,831	Non-endospermic seed	670	10	280
Broad bean	*Vicia faba* L.	4,332	Non-endospermic seed	560	10	230
Pigeon pea	*Cajanus cajan* L. Millsp.	4,454	Non-endospermic seed	560	20	250

Continued

Table 1.1. Continued.

Crop		World production[1] (1000 t)	Harvested unit	Seed composition[2]		
				Carbohydrate (g kg^{-1})	Oil (g kg^{-1})	Protein (g kg^{-1})
Others[9]						
Rapeseed[10]	*Brassica napus* L., *B campestris* L.	67,789	Non-endospermic seed	190	480	210
Sunflower	*Helianthus annuus* L.	40,931	Cypsela	480	290	200
Sesame	*Sesamum indicum* L.	4,738	Non-endospermic seed	190	540	200
Safflower	*Carthamus tinctoris* L.	776	Cypsela	500	330	140

[1]Average of 2011 to 2014, FAOSTAT (2016). [2]Seed composition data from Bewley *et al.* (2013), Sinclair and de Wit (1975), Langer and Hill (1991), Hulse *et al.* (1980), and Alberta Agriculture and Forestry (2015). [3]*Triticum aestivum* L. most common. [4]Harvested grain usually includes the lemma and palea. [5]May include members of other genera such as *Pennisetum, Papspalm, Setoria* and *Echinochla*. [6]In the shell. [7]Also includes other species of *Phaseolus* and, in some countries, *Vigna* species. [8]May include *P. arvense* (field pea). [9]Rapeseed is in the *Brassicaceae*, sunflower and safflower are in the *Asteraceae*, and sesame is in *Pedaliaceae*. [10]May include industrial and edible (canola) types, data from some countries includes mustard (*Brassica juncea* (L.) Czern, et Coss).

not be 'better' species waiting to be discovered. Nearly 100 years of intensive plant breeding produced the high-yielding cultivars of today's common crops; the need for a time investment of this magnitude in a new crop is a serious impediment to its successful deployment.

The harvested seed is a caryopsis in nine of the 22 species in Table 1.1, including the major crops maize, rice and wheat. Nine of the 22 species produce non-endospermic seeds; prominent crops in this group include soybean, groundnut and bean.

Composition of the seeds of these species varies widely (Table 1.1). Nine species, the cereals, produce seeds that are high in starch (>600 g kg$^{-1)}$ and low in protein (≤ 131 g kg^{-1}). Seeds of the traditional pulse or legume crops (seven species – bean, chickpea, dry pea, cowpea, lentil, broadbean and pigeon pea) have relatively high concentrations of protein (≥ 230 g kg$^{-1)}$, high to intermediate carbohydrate levels, and very low oil concentrations. Four species (rapeseed (canola), sunflower, sesame and safflower) are classified as oil crops, with high concentrations of oil (290–540 g kg^{-1}) and relatively high protein levels, with safflower a conspicuous exception (Table 1.1). Soybean and groundnut fall into a class by themselves, with seeds that contain exceptionally high protein (310–370 g kg^{-1}) concentrations and moderate (170 g kg^{-1}, soybean) to high (480 g kg^{-1}, groundnut) oil concentrations.

The seeds that sustain humankind were selected over the millennia from an enormous number of potential crop species. The grass seeds, the staff of life, are major sources of carbohydrates for much of the world and are complemented by the pulses (legumes) with their relatively high protein levels (poor man's meat) (Heiser, 1973, p. 116). These crops have fed humankind for centuries and it seems likely that we will continue to rely on them for the foreseeable future. Fortunately, the productivity of these crops has increased in step with the expanding world population.

Increasing Food Supplies: Historical Trends in Seed Yield

World population has increased by approximately 1000 times since the beginning of agriculture (Cohen, 1995, p. 30). The world population was roughly one billion (Cohen, 1995, p. 400) at the turn of the 19th century, when Thomas Malthus made his apocalyptic prediction (1798) that the power of population to increase is indefinitely greater than the power of the earth to provide food. The world population reached 7.3 billion in 2015, accompanied by food supplies that are, overall, more than adequate, as indicated by low grain prices in many countries, record low levels of undernourished people and rising concerns of an obesity epidemic in developed countries (FAOSTAT, 2016). Food supplies have increased since Malthus's day more or less in step with population.

There are only six basic avenues by which food production can be increased (Evans, 1998, p. 197).

1. Increase the land area under cultivation
2. Increase the crop yield per unit area
3. Increase the number of crops per unit area per year (multiple cropping)
4. Replace lower yielding crops with higher yielding crops
5. Reduction of post-harvest losses
6. Reduced use as feed for animals.

The first four options deal with the quantity of food produced by crops, our interest in this book, but the last two would also increase the amount of food available for consumption by the world's population. Shortening the food chain by utilizing more plant and fewer animal products, and reducing waste in harvest, storage and utilization of food and feedstuffs could make significant contributions, as could reducing the land area devoted to non-food production (i.e. crops fed to cats, dogs, horses and other pets; fibre, industrial, and especially biofuel crops). All of these last options would contribute to a larger food supply without increasing the land used for crop production, yield per unit area or the inputs required to increase yield. We will come back to these non-production options in Chapter 6, but they all involve complicated economic and social issues that are mostly beyond the purview of crop physiologists and this book.

Historical increases in food production were often associated with cultivation of more land. For example, wheat and maize production in the US increased by 3.5- to fivefold from 1866 to 1920 as a result of a three- to fourfold increase in harvested area as production moved west onto new lands in the Corn Belt and Great Plains states (NASS, 2016). The shift from the use of animal power (primarily horses and mules) to mechanical power (cars, tractors, trucks) fuelled by petroleum products in the early years of the 20th century reduced the need for feed production and made more land available for food production. Increases in yield, however, played a much larger role in more recent times as the supply of unused land declined.

Yield from eras closer to the beginning of agriculture 10,000 years ago provide an interesting perspective on current discussions of yield and the potential for yield improvement. Estimated maize yields in Mexico in 3000 BC were approximately 100 kg ha^{-1}, while brown rice yields in Japan in 800 AD were 1000 kg ha^{-1} (Evans, 1993, pp. 276–279). Wheat yield in England increased from roughly 500 kg ha^{-1} in 1200–1400 AD to approximately 1100 kg ha^{-1} in the 1700s and nearly 2000 kg ha^{-1} in the 1800s (Stanhill, 1976). Wheat yields in New York averaged 1077 kg ha^{-1} for the period from 1865–1875 (Jensen, 1978). Modern yields (2011–2014 averages) for comparison are 7593 and 4182 kg ha^{-1} for wheat in England and New York, respectively; 6707 kg ha^{-1} for rice in Japan; and 3146 and 9391 kg ha^{-1} for maize in Mexico and the USA (FAOSTAT, 2016; NASS, 2016). Clearly yields have increased along with the world's population.

Documentation of changes in crop yield over a shorter time frame in the USA is shown in Fig. 1.1 for two cereals (maize and wheat) and a legume (soybean). There was relatively little change in yield of maize and wheat from 1866 to ~1940, when the advent of high-input agriculture (chemical fertilizers, herbicides

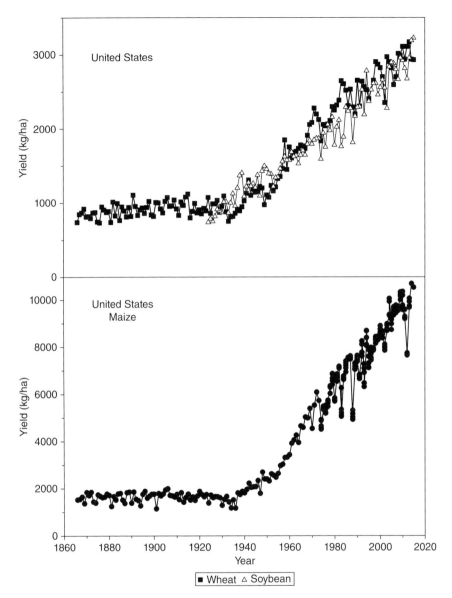

Fig. 1.1. Average yields of maize, wheat and soybean in the United States. Data from the National Agriculture Statistics Service (NASS, 2016).

and pesticides) combined with the use of hybridization to produce improved cultivars (hybrids in maize, but not wheat) started a steady increase in yield that has continued to the present time. Soybean yield in the USA also increased steadily from 1924; the first year that yield data were available. The three- to sixfold increases in yield of these crops in the 75 years after 1940 is truly astounding when compared with the previous 74 years, when there was no change. The agricultural systems in place for that 74-year period were low-input systems that emphasized a mixture of crop and animal agriculture and multi-crop rotations that included legumes with manure providing much of the N input (Egli, 2008); a system that would probably fit the modern day definition of organic agriculture.

World yields of wheat, maize and rice (Fig. 1.2) also increased steadily from 1961 to 2012. World yields from earlier years are not readily available, but they probably followed a pattern similar to those in Fig. 1.1.

Any evaluation of historical yield trends leads to the question – what will happen in the future? Will the increase continue indefinitely (surely there is a maximum set by biophysical limits on the conversion of solar energy to biomass) or will it slow and eventually stop, resulting in a yield plateau? There is no clear evidence in Fig. 1.1 and 1.2 that yields are reaching a plateau. There is, however, evidence for plateaus in some crops in some production systems (e.g. wheat in France (Brisson *et al.*, 2010), rice in Korea and China, wheat in northwest Europe and India, and maize in China (Cassman *et al.*, 2010)). It is very difficult to iden-tify yield plateaus, and many apparent plateaus in the past were only temporary cessations in yield growth. In the first edition of this book (Egli, 1998, pp. 6–7), US and world wheat yields exhibited plateaus for the last 14 (USA, 1983 to 1996) and six (world wheat, 1990 to 1995) years of record, but Figs 1.1 and 1.2 show

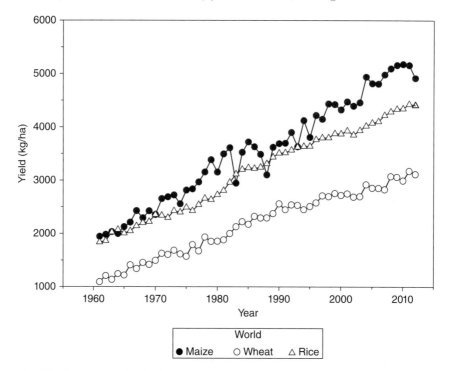

Fig. 1.2. Average world yields of maize, wheat, and rice, 1961 to 2014. Data from FAOSTAT (2016).

that these were only temporary plateaus, and yield eventually resumed its upward trend. It is always possible in any yield time series to identify short periods when there is no yield growth, but then growth begins anew and the plateau disappears. Rigorous statistical protocols to detect yield plateaus have been developed (Lin and Huybers, 2012; Grassini *et al.*, 2013), but statistical analysis cannot predict future yields and it is those yields that determine whether a plateau persists or the increase in yield resumes. Plateaus are often a result of sub-optimal environmental conditions, but they may also reflect a lack of production inputs, government policy, or emphasis on quality over yield (Fischer *et al.*, 2014, pp. 41–43) and do not always reflect fundamental limitations of the plant. Yield plateaus will seriously limit our ability to maintain adequate food supplies for an increasing world population, so the question – how long and how rapidly will yields continue to increase? – is extremely important. We will return to these issues in Chapter 6.

The steadily increasing yields in Figs 1.1 and 1.2 were primarily the result of two basic changes. Either the plant was improved through plant breeding and selection, or the plant's environment was improved by crop management. Improvements from breeding are frequently divided into those increasing yield via defect elimination and those increasing yield in a non-stress environment (potential yield) (Donald, 1968). Defect elimination allows the farmer to 'recover' the yield that would have occurred in the absence of the defect, but does not add to the potential yield. An example of defect elimination was reported by Sandfaer and Haahr (1975) where the yield of old cultivars of barley was 26% lower than new cultivars when the evaluations were made in the presence of the barley yellow stripe virus but only 8% lower in the absence of the virus. Much of the higher yields of the new cultivars came from incorporation of virus resistance, i.e. elimination of a defect (susceptibility to the virus), and not through any change in the primary productivity of the plant. Both approaches contribute to higher yield in the farmer's field, but the relative contribution of the two is not well defined and no doubt varies among crops and cropping systems.

Both breeding and management contributed to past increases in yield and, in many cases, new cultivars were only effective when management practices changed. For example, the shorter rice cultivars that were at the heart of the green revolution produced higher yields only when they received high levels of N fertilizer (Chandler, 1969); modern maize hybrids express their superior yielding ability only when grown at high population densities (Duvick, 1984).

The traits that Duvick (1992) associated with higher yielding maize hybrids included defensive traits (i.e. defect elimination) such as resistance to premature death, stalk and root lodging resistance, shorter anthesis–silking intervals resulting in less barrenness, and tolerance to European corn borer (*Ostrinia nubilalis* Hubner). More upright leaves (probably contributing to higher canopy photosynthesis) and longer seed-filling periods (Cavalieri and Smith, 1985) probably represent direct selection for potential yield. Increasing the harvest index, the ratio of yield to total biomass, was associated with improvement in potential yield of wheat, barley (Evans, 1993, pp. 238–260) and rice (Peng *et al.*, 2000) with no change in total biomass, although more recent evidence suggests that increases are now driven by increases in total biomass (Peng *et al.*, 2000; Shearman *et al.*, 2005). Changes in many other plant characteristics have been related to improvement of potential yield and defect elimination (see Evans, 1993, pp. 169–268 for a thorough discussion of this topic), but it is not always clear that these historical changes provide any guidance for future improvements.

Estimates of the proportion of the total yield increase coming from plant breeding range from 20 to 80% across several crops (Evans, 1993, pp. 297–307). Estimates for some of the major grain crops (maize, wheat, soybean, sorghum) in the USA suggest that from 40 to 80% of the yield increase came from plant breeding (Smith *et al.*, 2014; Schmidt, 1984; Specht *et al.*, 2014; Miller and Kebede, 1984). The total breeding effort, breeding objectives, and the quality of the environment influence progress from breeding (Evans, 1993, p. 307), so relatively low yields of some minor crops (i.e. crops grown on limited acreage, such as some grain legumes) may partially reflect limited breeding efforts. Precise estimates of the relative contributions of breeding and management are difficult and probably vary widely among crops and cropping systems. The contribution from crop management, however, will probably decrease in the future, as past improvements make the next increment in yield more difficult (Egli, 2008).

What will happen in the future is a much-debated question, a debate that focuses on three major topics with very little agreement on any of them. The three main issues are: (1) Will yields keep increasing and will the increase be adequate to feed an expanding, more affluent population? This yield question is particularly important because expansion of the land area used to produce food is usually considered an undesirable approach. (2) What effect will global climate change have on production – will reductions in production from higher temperatures and lower rainfall exceed gains from higher rainfall or from expansion of crop lands to areas where production is not currently possible (e.g. current expansion of maize production into the prairie provinces of Canada)? (3) What effect will shifting from coal and petroleum to energy sources that emit fewer greenhouse gases, such as solar and wind, have on agricultural productivity? Much of the increase in agricultural productivity in the high-input era was based on cheap energy, raising the question: Can productivity be sustained and increased with more expensive energy? These are all complex questions, and the hopes and fears they raise will be discussed in more detail in Chapter 6.

Crop Physiology and Yield Improvement

Plant growth and the production of yield can be studied at varying levels of complexity, from the plant community to the molecular level, i.e. crop community, plant, organs, tissues, organelles, macromolecules and atoms/molecules (Thornley, 1980). Economic yield of grain crops, however, is always measured on a land area basis and must be studied as a community phenomenon, not as the product of individual plants. Consequently, agronomists have traditionally evaluated yield at the community level. Many factors that they consider to be important, such as plant population, leaf area index and solar radiation interception, are characteristics of a community of plants, not individual plants; other important factors may be characteristics of the individual plant (e.g. C_3 or C_4 photosynthesis, leaf display). Characteristics that make an isolated plant productive may have no effect or a negative effect at the community level. Leaf angle is a classic example of this phenomenon; isolated plants benefit from horizontal leaves, while community productivity may be higher with a mix of horizontal and vertical leaves (Duncan, 1971).

Scientific investigations of plant growth go back at least to the work of Priestle in 1771 (plants released oxygen); Ingenhouse (light required for the evolution of oxygen by plants) and de Saussine, who showed, in 1804, that plants took up mineral nutrients and NO_3 from the soil (Evans, 1975, p. 12). Crop physiology, understanding the dynamics of yield production of crops, began with the work of W. L. Balls in the early 1900s on plant spacing and sowing dates with cotton (*Gossypium* spp.) communities in Egypt, *not* isolated plants (my emphasis) (Evans, 1975, pp. 13–14).

Growth analysis techniques were developed in the first half of the 20th century to describe growth of plants and plant communities (Blackman, 1919; Watson, 1947, 1958). The components of growth analysis describe the accumulation of dry matter with a general goal of learning more about plant or community characteristics that regulate productivity. The absolute growth rate (g plant^{-1} day^{-1} or g m^{-2} $_{land\ area}$ d^{-1}) provides the starting point and other growth analysis parameters deconstruct the absolute rate to better understand its regulation.

The relative growth rate (RGR, g g$_{dry\ weight}$ $^{-1}$, Blackman, 1919) describes the inherent ability of the plant to accumulate dry matter per unit of dry matter present. Photosynthesis by leaves is responsible for almost all of the dry matter accumulation by crop plants, so expressing dry matter accumulation on a leaf area basis, i.e. net assimilation rate (NAR, g m^{-2} $_{leaf\ area}$ day^{-1}, Briggs *et al.*, 1920), provides a better representation of growth capacity than RGR based on total plant weight. Since leaves are the primary source of photosynthesis, the proportion of dry weight allocated to leaves, the leaf area ratio (LAR, m^2 leaf area g $_{dry\ weight}$ $^{-1}$, Briggs *et al.*, 1920) is also an important parameter.

The absolute rate of accumulation of dry matter by a crop community, the crop growth rate (CGR) expressed as g m^{-2} $_{land\ area}$ day^{-1} always refers to the growth of the crop community, never to growth of individual isolated plants. Watson (1947) defined leaf area index (LAI), the ratio of leaf area (one side only) to the ground area, as a convenient way of describing the leaf area of a crop. An LAI of 2 means that there are 2 m^2 leaf area per m^2 ground area. The leaf area duration (Watson, 1947) interjects time into the analysis by considering how long the leaf is present. The CGR is, in its simplest form, determined by the amount of intercepted solar radiation (a function of LAI and leaf display) and its conversion by the plant into dry matter (radiation use efficiency, dry matter per unit intercepted radiation, g MJ^{-1} (Wilson, 1967), as shown in equation 1.1:

Where $\text{CGR} = (\text{SR})(\text{SRI})(\text{RUE})$ (1.1)

$\quad$ CGR = crop growth rate $(\text{g m}^{-2}\text{day}^{-1})$

$\quad$ SR = daily incident solar radiation $(\text{MJ m}^{-2}\text{day}^{-1})$

$\quad$ SRI = proportion of SR intercepted by the plant community (%) and

$\quad$ RUE = radiation use efficiency (g MJ^{-1}).

Growth analysis techniques provide a simple framework to help us understand the basis for differences in the absolute growth rate and productivity of individual plants or plant communities. Hunt (1978) provides a detailed summary of growth analysis techniques.

$\quad$ The growth analysis approach is useful because it highlights important plant and community characteristics that control productivity. The growth analysis equations remind us that differences in biomass can result from variation in simple plant or community characteristics and are not always dependent upon the inherent metabolic ability of the plant. The production of leaves to intercept solar radiation, a function of LAI, leaf area ratio, plant density and special arrangement of plants, is a key to determining CGR, so substantial differences in CGR could be completely independent of the inherent photosynthetic capacity of the plant. Variation in the growth rate of seedlings may be related to the size of the planting seed which determines the initial leaf area, solar radiation interception and the absolute growth rate without any differences in the inherent productivity (Egli *et al.*, 1990). Higher leaf area ratios will also accelerate the growth of isolated seedlings.

$\quad$ Crop physiologists too often emphasize metabolic aspects of growth and ignore simpler characteristics, even though they are clearly identified by growth analysis techniques. The growth analysis approach clearly differentiates between isolated plants and plant communities, a distinction that is often ignored by fundamental plant scientists. For example, large plants with many leaves and a large LAI may grow faster and yield more in isolation, but show no advantage over smaller plants in a community setting. Intercepted solar radiation (equation 1.1) of isolated plants is directly related to LAI, however, in a community solar radiation interception increases with LAI until it approaches 100% (complete ground cover); increasing LAI above this level will not increase intercepted solar radiation or CGR. A plant that produces many tillers or branches performs well as an isolated plant, but loses its advantage in a community because the extra LAI associated with the tillers or branches does not increase solar radiation interception.

$\quad$ Although growth analysis techniques provide a useful description of plant growth and made significant contributions to our understanding of the basic processes involved, they have a number of weaknesses that limit their usefulness. Measurements of plant dry weight are typically quite variable, especially in the field, which reduces the precision of parameter estimates and the ability to detect treatment effects. This lack of precision limits meaningful estimates of growth analysis parameters over short intervals, while average values from samples taken at weekly or greater intervals do not provide much information about short-term environmental effects on growth.

Some growth analysis concepts, especially NAR and RGR, do not provide useful information when applied to plant communities. Once solar radiation interception by the community reaches a maximum, CGR is constant (ignoring environmental effects) (Shibles and Weber, 1965), but plant weight and LAI continue to increase. A constant growth rate combined with increasing plant weight and LAI cause RGR (growth rate per unit dry weight) and NAR (growth rate per unit leaf area) to decline. These declining rates do not provide useful information about crop growth.

The original growth analysis formulations did not deal explicitly with reproductive growth, which limited their application to understanding the production of grain yield. This deficiency was later remedied by the work of Wilson (1967) and Charles-Edwards (1982), and the development of the harvest index concept (Donald, 1962).

In spite of the limitations of growth analysis approaches, they provide a useful theoretical framework to guide our thinking about crop productivity. These concepts should not be forgotten in the current high-tech crop physiology research environment. In fact, the vestiges of growth analysis can be found in many current descriptions of crop growth, including the widespread use of CGR and radiation use efficiency.

In the middle of the 20th century, physiologists began to shift their emphasis to lower levels of complexity, to the organ level or below (Boote and Sinclair, 2006), as they investigated basic plant growth processes such as photosynthesis, nitrogen fixation, nitrate reduction and assimilate transport. This shift was probably partially driven by the inability of growth analysis techniques to address more fundamental questions about plant growth raised by a deeper understanding of plant metabolism. The availability of simple infra-red gas analysers to measure CO_2 concentrations opened the door to extensive study of single-leaf (Hesketh and Moss, 1963) and canopy photosynthesis (Larson *et al.*, 1981). The underlying assumption of this approach was that studying the fundamental metabolic processes involved in plant growth would lead to a better understanding of the yield production process. It often proved difficult, however, to relate information about the basic functioning of a process to the growth of an intact plant or a plant community.

In theory it should be possible to integrate information across all levels, from the molecular level to the plant community but this has proven to be difficult and may be practically impossible (Thornley, 1980; Sinclair and Purcell, 2005). Even using knowledge of the biochemistry of plant processes to predict canopy photosynthesis or CGR seems beyond the realm of possibility. The problem may be one of complexity; crop growth and yield are the end result of many individual plant processes and cycles operating over time, making it difficult to integrate knowledge of them together in a useful fashion. Some would argue that not enough is known about the processes to put them together; more research is needed and then yield can be explained, starting at the molecular level. Another possibility may be that the usefulness of information of processes at lower levels is limited by the dominance of whole plant–plant community characteristics in determining yield.

W.G. Duncan, one of the original crop modellers, addressed this dilemma when he described the study of the pieces of the photosynthetic apparatus as 'something like being given the pieces of a good watch in a box and then being asked what time it is' (Duncan, 1967, p. 309). Duncan was making the point that basic knowledge of an individual process, in this case photosynthesis (i.e. light reaction, Rubisco, etc.) does not necessarily provide any useful information about the functioning of the plant community, i.e. canopy photosynthesis and the production of yield.

The focus on basic plant growth processes was followed by renewed interest at the whole plant–plant community level (Boote and Sinclair, 2006), which may have reflected our inability to integrate knowledge from lower levels to the whole plant or plant community level. Current research has again shifted to lower levels (Boote and Sinclair, 2006), probably driven by developments in molecular biology with its focus on specific genes and their role in regulating plant growth. Boote and Sinclair (2006) suggested that this cycling between a narrow focus at the gene level and whole plant and plant community studies will continue in the future. This cycling may eventually blur the difference between basic knowledge and its significance in the yield production process.

The complexity of the yield production system and the inability to integrate knowledge from basic levels to the functioning of the plant community stimulated interest in the development of crop simulation models. These models were visualized as tools to understand how the bits and pieces of the system contributed to the functioning of the community. The first models took a very simplistic approach to crop growth; for example, one of the first models (Duncan *et al.*, 1967) simply calculated the daily photosynthesis of a crop community as a function of photosynthetic system (C_4 or C_3), leaf area, leaf display (leaf angle) solar radiation and a solar radiation–single leaf photosynthesis response curve. One of the contributions of this simple model was to quantify the effect of leaf angle on canopy photosynthesis, a relationship that was much debated at that time, and to show that vertical leaves only increased canopy photosynthesis at high LAIs (Duncan, 1971). These findings illustrate one of the key functions of a model – the ability to evaluate relationships that are very difficult to test experimentally (Boote *et al.*, 1996). De Wit (1965) also made significant contributions to the early development of crop simulation models and, from those early beginnings, the models developed to the point where they 'grow' crops from planting to maturity. These models eventually included water relations, mineral nutrients, respiration, partitioning, and temperature effects and produced estimates of yield often expressed as the number of seeds per unit area and seed size (weight per seed). Some models were included in a systems package (e.g. the DSSAT family of models, Jones *et al.*, 2003) that made it possible to conduct multi-year comparisons of various management strategies; in short, they were sophisticated tools for studying management and environmental effects on crop growth and yield. In recent years, crop simulation models provided insights into the potential effects of climate change on crop yield (Asseng *et al.*, 2009), insight that would be very difficult to obtain experimentally.

Crop simulation models made contributions to our understanding of the yield production process, but I don't think they had the impact envisioned by the early pioneer modellers. Models have rarely contributed great insights into the fundamental processes controlling grain yield. The ability to manipulate individual processes and relationships with no limitations would seem to be a crop physiologist's dream, but it hasn't been as useful as expected. In spite of the ability of models to evaluate the effect of management practices on yield for multiple locations and years, applied agronomists continue to laboriously evaluate the same practices in field experiments year after year. Models would seem to be the perfect adjunct to the development of precision agriculture practices, but again they seem to have had only marginal impacts.

One limitation to the use of crop simulation models is that they are still too simplistic to capture all important aspects of the yield production process. A simplistic representation of a complicated process does not necessarily provide a strong basis for in-depth investigations of that process. I think the impact of crop models is also limited by a lack of interaction between crop physiologists (experimenters) and modellers. Crop physiologists designing experiments to answer questions raised by modellers, and modellers testing hypotheses to sharpen the focus of crop physiologists (Passioura, 1996) has not, in my opinion, occurred on a wide scale, certainly to a lesser degree than the interactions between theoreticians (equating a crop simulation model to a theoretical description of crop growth) and experimenters in other disciplines. This interaction and the entire modelling endeavour may have been limited by the absence of funding streams for the explicit development of models to study yield production in grain crops.

We now have a much better understanding of how crop plants grow and produce yield, thanks to the efforts of crop physiologists, other plant scientists, and modellers, than we had in the middle of the last century when yields started their rapid increase (Figs 1.1, 1.2). Our understanding of the yield production process will, no doubt, continue to improve; the challenge is to use this understanding to improve our crop production systems in the face of an uncertain future.

The Seed: an Integral Component of the Yield Production Process

A fairly detailed understanding of crop growth and the production of yield is now available at the community level. Crop physiologists and modellers, however, have been slow to consider the seed as an explicit component of the system, but the seed cannot be ignored because only the dry matter accumulated by the seed is harvested for yield. It is worth noting that vegetative growth, before the seeds start accumulating dry matter, is just a preliminary activity; at the beginning of seed growth, no yield has been produced, it is all produced during the seed-filling period. Granted, leaves, stems and roots provide the synthetic capacity to feed the seeds, but all storage materials that give seeds their value (oil, protein, and starch) are synthesized largely in the seed from raw materials produced in the leaves. This synthetic capacity makes the seed a critical component of the yield production system in grain crops. Consequently, including the seed in the yield production process will lead to a more complete understanding of the system.

The seed has a dual function in agronomic crops, it serves, as planting seed, to regenerate the crop, and it is the organ harvested for economic yield. Of course, the growth and development of the seed on the plant are the same if the ultimate fate of the seed is to be planted in the soil to produce the next crop or if it is to be eaten or processed for food, feed or industrial purposes. The two seeds, planting seed or grain, however, are not equal from a crop management viewpoint. The attributes of quality are not the same and consequently the management practices for producing high-quality planting seed are not always the same as those used to produce seed for grain. Planting seed must be genetically pure, viable and vigorous; traits not important for seed produced for grain. My focus in this book is on grain yield and the role that the seed plays in determining yield. I will not consider the essential role of the seed as a regenerator of the crop because this role has been covered at length by other authors (e.g. McDonald and Copeland, 1997; Copeland and McDonald, 2001; Bewley *et al.*, 2013).

Many formulations of the production of yield describe the accumulation of dry matter by a crop community and then simply partition or allocate some portion of this dry matter to the harvested fraction, for our purposes, the seed (Wilson, 1967; Charles-Edwards, 1982; Sinclair, 1986). This approach emphasized that yield was not solely a function of the ability of the crop to accumulate dry matter, but also a function of how much dry matter was allocated to the reproductive fraction. Unfortunately, this allocation was represented by a simple ratio at maturity that did not provide any mechanistic insights into the yield production process.

Growth analysis techniques emphasized understanding the processes involved in the production of dry matter and largely ignored the processes regulating the accumulation of dry matter by the seed. Division of yield production into the production of assimilates by the source and utilization of those assimilates by the sink included seeds in the evaluation, but the sink (seed) was too often assumed to be a simple receptacle for assimilate produced by the leaves. The seed was directly involved in investigations of yield components – plants per unit area, pods per plant, seeds per pod and weight per seed for a grain legume. Relationships among these components were studied to learn more about how the plant produced seed yield. Much of this research, however, represented a statistical search for relationships among components and contributed little to our understanding of yield production. Yield component compensation – when changes in one component were frequently associated with changes in the opposite direction in another component with no change in yield – gave yield components a bad reputation. I will attempt a fresh look at yield components in later chapters that will, hopefully, improve their reputation.

Although past investigations that included the reproductive fraction of the plant have not been particularly useful, it is my thesis that the processes involved in determining the proportion of the total biomass that ends up in the seed, i.e. grain yield, cannot be understood at a mechanistic level without considering the growth and development of the seed. Accumulation of dry matter by the plant community is the fundamental basis of crop yield, but it is not the only important process. The ability of the individual seed to accumulate dry matter is also important; after all, it's the seed that is harvested for yield and it should not be surprising that the ability of the seed to accumulate dry matter is an important consideration in understanding the yield production process. I believe the key to understanding many important yield formative processes (determination of seeds per unit area, seed size, source–sink relationships) is to include the characteristics of the seed in the analysis.

My objective in this book is to consider the production of yield by grain crops from the perspective of the individual seed. This will be accomplished by investigating the characteristics of growth and development of the individual seed, the regulation of growth and development and the influence of the environment and plant characteristics on growth and development. This information will be used to develop a mechanistic understanding of the role of the seed in the production of yield by grain crops.

My focus in this book will be primarily at the level of the organ, plant and plant community. I will not investigate seed growth at lower levels; the extensive information on the physiology and biochemistry of the processes underlying seed growth and the potential involvement of hormones will not be covered. There are two reasons for these omissions. First, these topics are already covered in great detail in other publications (e.g. Bewley *et al.*, 2013), so no particular purpose would be served by repeating that information here. Second, and perhaps more importantly, these topics, in my opinion, provide little useful information about the role of the seed in the determination of crop yield.

When one considers the great diversity in shape, colour, size and composition of seeds harvested from grain crops, the objective of this book may seem hopelessly ambitious, requiring, at best, several volumes. Fortunately, this is not so, because, as we shall see, the important characteristics and general patterns of seed growth are remarkably uniform across the species listed in Table 1.1, and perhaps across most plant species bearing orthodox (non-recalcitrant) seeds. This uniformity will make it possible to develop concepts describing the role of the seed in the production of yield that will apply to all grain crops.

Seed Growth and Development **2**

Seed Structure, Composition and Size

Seeds of grain crops seem to the novice to be quite variable because they exhibit large differences in size, shape and colour, but, at a more fundamental level of structure and function, there is much less variation. Most of the seeds harvested for food or feed come from species of only two families, the *Poaceae* (grasses) and *Fabaceae* (legumes), which limits the variation in seed characteristics (18 of 22 species, Table 1.1). This concentration in two families also limits variation in seed composition, with seeds from *Poaceae* uniformly high in starch and the non-endospermic seeds of the *Fabaceae* important sources of protein. Crops with high oil concentrations in their seeds (rapeseed, sunflower, sesame and sunflower) come from several other families (Table 1.1).

The composition of oil and protein in the seeds also varies among and within species and this variation plays an important role in determining quality and economic value of the products produced from these seeds. Current interest in healthier foods favours some species or cultivars over others (e.g. rapeseed cultivars that produce edible oils over traditional oil sources, such as soybean, that are higher in saturated fats) and stimulated development of cultivars with more desirable oil profiles. To date, commercialization of these cultivars has often proved difficult. Synthesis of oil and protein requires more metabolic energy than synthesis of starch (Penning de Vries *et al.*, 1974), thus seed composition affects potential yield, which explains some species differences in yield. Variation in energy requirements also explains why genetic manipulation of seed composition can affect yield, as shown, for example, by the yield reduction that often occurs when plant breeders increase seed protein concentration (Brim and Burton, 1979). Seeds exhibit tremendous variation in size (weight per seed), ranging from 0.001 mg seed^{-1} (orchid) to 20 kg seed^{-1} (double coconut, *Lodoicea maldivica*) (Moles *et al.*, 2005). The variation among the species in Table 1.1 is less but still substantial, ranging from less than 10 mg seed^{-1} (millet, rapeseed and sesame) to more than 2000 mg seed^{-1} (broad bean) (Briaty *et al.*, 1969). Seeds of the *Poaceae* are usually relatively small (less than 50 mg seed^{-1}) with maize, whose seeds generally weigh more than 200 mg, representing an obvious exception. Legume seeds tend to be larger, with most species producing seeds in excess of 100 to 200 mg, with the exception of chickpea and lentil, whose seeds usually weigh less than 50 mg. Seeds of the oil crops vary from small (< 10 mg seed^{-1}) for rapeseed and sesame to intermediate (30–70 mg seed^{-1}) for sunflower and safflower. More information on species differences in seed size can be found in Table 3.1.

The caryopsis of the *Poaceae* is a fruit, not a seed, because the pericarp is fused to a rudimentary testa surrounding the endosperm and embryo (Fig. 2.1a, c). In this book, I will follow the generally accepted practice of referring to the fruit of the *Poaceae* as a seed. The well developed starchy endosperm, with an outer aleurone layer, can comprise as much as 80% of the dry weight of the mature seed. The mature endosperm, except for the aleurone layer, consists of dead cells packed with starch and some protein. The embryo is relatively small, accounting for only about 10% of the seed dry weight. The single cotyledon has been modified to form the scutellum, which is a rich source of oil in some species. The embryo is usually located on one side of the seed near the point of attachment of the seed to the mother plant (Fig. 2.1c).

There is no vascular connection between the maternal tissue and the embryo or the endosperm. Consequently, assimilate must unload from the phloem in the maternal tissues and move apoplastically into the embryo and endosperm before being taken up by the cells. In maize, the unloading occurs in the pedicel tissue, the tissue connecting the seed to the cob (Fig 2.1c). Assimilates move apoplastically through the placenta-chalaza region to the endosperm. Movement into the endosperm may be facilitated by endosperm transfer cells located at the boundary between the placenta-chalaza regions (Thorne, 1985).

In wheat and barley, phloem unloading occurs in a single vascular bundle, embedded in the maternal tissue and running the length of the kernel at the bottom of the crease (Fig. 2.1a). Rice has a single vascular bundle embedded in the pericarp and unloading occurs along the entire length of the seed (Oparka and Gates, 1981). In all cases, after unloading from the phloem, assimilates move apoplastically into the embryo and endosperm and are taken up by the cells.

The non-endospermic true seed of the *Fabaceae* consists of a large embryo surrounded by the testa or seed coat (Fig. 2.1b). The embryo consists of two large cotyledons and the embryo axis. The majority of the reserve materials are stored in the cotyledons which make up as much as 90% of the total seed dry weight.

Assimilate moves through the funiculus into the testa from the vascular bundle on the ventral suture of the pod. A single vascular bundle enters the testa at the chalazal end of the hilum and spreads throughout the testa with the vascular patterns varying among species from one or two phloem strands in *Vicia* and *Pisum* to reticulate venation in *Glycine* and *Phaseolus* (Thorne, 1985). There is no vascular connection between the testa and the embryo, thus assimilate moving into the seed must be unloaded from the phloem in the testa and moved apoplastically into the embryo where it is taken up by cotyledon cells. A more detailed description of seed structure can be found in Bewley *et al.* (2013, pp. 1–7) while the movement of assimilate into the seed was reviewed by Patrick and Offler (2001).

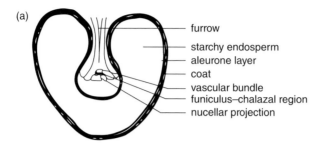

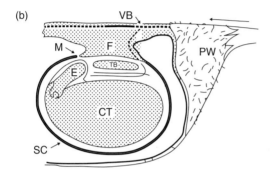

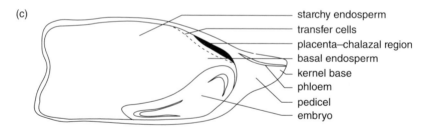

Fig. 2.1. (a) Cross-section through a developing wheat seed at mid-point between apex and base to show the relationship between the vascular tissue and the starchy endosperm. Adapted from Bewley and Black (1994).(b) Sketch of a typical median sagittal section of an entire soybean seed attached by the funiculus to a ventral bundle of the pod. A single vascular bundle enters the seed coat at the chalazal end of the hilum and branches below the tracheid bar to form two lateral bundles. CT, cotyledon; E, embryonic axis; F, funiculus; M, micropyle; PW, pod wall; SC, seed coat; TB, tracheid bar; VB, vascular bundles. Adapted from Thorne (1981).(c) Longitudinal section through a developing maize seed. Adapted from Bewley and Black (1994).

Although there is variation in structure, composition and size among seeds of the 22 plant species that provide much of our food supply, there is enough uniformity in growth characteristics to develop a general description of the growth and development of seeds that applies to all species in Table 1.1. For example, all of the species in Table 1.1 produce orthodox (non-recalcitrant) seeds by sexual reproduction and, in all cases, there is no vascular connection between the embryo, representing the next generation, the endosperm and the mother plant. Growth of all of these seeds requires transport of assimilate and water across this discontinuity. As we investigate the general patterns of growth and development, we will encounter other aspects of seed growth that are uniform across species. In fact, the uniform characteristics will be of much greater importance in understanding seed growth and its role in the production of yield than characteristics exhibiting diversity among species. Seed size and structure, for example, are not important determinants of yield. The importance of the common characteristics makes it possible to develop a single unified description of the involvement of the seed in the production of yield.

The Three Phases of Seed Development

Seed development begins with the production of the flower primordia long before anthesis. The developing flower contains tissues that will ultimately be part of the fruit and seed. The pod walls (carpels) of the legume fruit and the pericarp of the cereal caryopsis develop from the ovary. The testa forms from the integuments around the ovule. Thus, the 'seed' that represents economic yield is a mixture of embryonic and maternal tissues. The mature seed could conceivably be influenced by developmental processes occurring before anthesis; however, I will restrict my discussion of seed development in this chapter to the period beginning at fertilization. Excellent coverage of seed development in much more detail can be found in Bewley *et al.* (2013, pp. 27–52).

Seed development, from fertilization to the mature seed, can be divided into three phases (Adams and Rinne, 1980). Phase I includes fertilization and the period of rapid cell division when all seed structures are formed. Phase II is when the seed accumulates reserve materials that give it economic value. Phase III begins when the accumulation of reserve materials slows prior to stopping at physiological maturity. The growth curve for an individual soybean seed illustrates these three phases (Fig. 2.2). The dry weight of an individual seed increases slowly during the initial lag phase (phase I) and then it increases rapidly to a constant maximum rate during the linear phase (phase II), after which the growth rate decreases to zero at physiological maturity (maximum seed dry weight, phase III). Seeds of wheat (Fig. 2.3), maize (Fig. 2.4) and all other grain crop species follow the same pattern of dry matter accumulation.

Water concentration (g H_2O per g fresh weight) is very high during phase I and declines steadily until the seed reaches physiological maturity. For example, the water concentration of soybean seed is above 800 g kg^{-1} (80%) early in

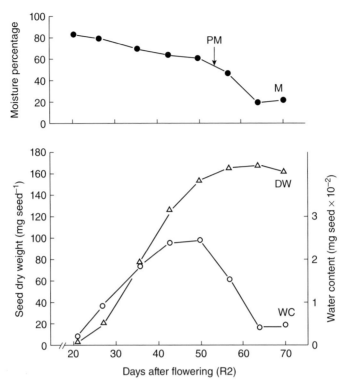

Fig. 2.2. Seed dry weight (DW), water content per seed (WC) and water concentration (M) of an individual soybean seed developing in a field environment. PM, physiological maturity. From Fraser *et al.* (1982a).

development (Fig. 2.2) and declines steadily to about 550 g kg^{-1} (55%) at physiological maturity. Similar patterns have been reported for wheat (Fig. 2.3) and maize (Fig. 2.4 and Westgate, 1994), although the concentration at physiological maturity varies significantly among species (Fig. 2.5, Table 2.1). There is also some variation among cultivars within a species (Table 2.1); the significance of this variation will be discussed in Chapter 3.

The water potential of seeds or seed tissues does not change much during phase II of seed development as shown by the water potential of maize embryos which remained between −1.0 and −1.5 MPa during the linear phase of seed development (Fig. 2.4). Egli and TeKrony (1997) reported results for wheat embryos and soybean axes that were similar to those of Saab and Obendorf (1989) (soybean axes) and Westgate (1994) (maize embryos), although they reported slightly higher values (−0.5 to −1.5 MPa). Relatively constant water potentials suggest that the water status of the seed changes little during phase II of seed development and the large changes in seed water concentration are not indicative of the water status.

Soybean seed water content (mg per seed) increases during the early stages of development, reaches a plateau, and then declines rapidly after physiological

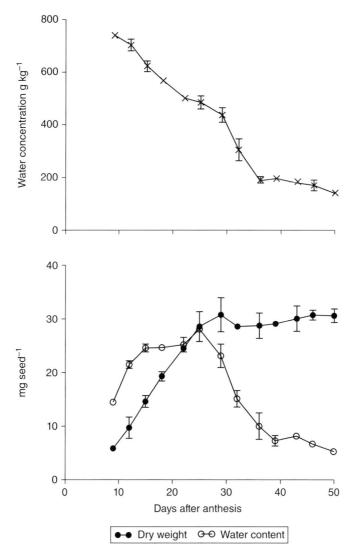

Fig. 2.3. Seed dry weight, water concentration (g kg⁻¹) and water content (mg seed⁻¹) of an individual soft red winter wheat seed developing in the field. From Ibrahim *et al.* (1992). Bars represent ± 1 standard deviation.

maturity (Fig. 2.2). Cereal seeds usually reach their maximum water content earlier in development, as shown for wheat (Fig. 2.3) and maize (Fig. 2.4; see also Gambin *et al.*, 2007). Maximum seed water content also represents maximum seed volume.

   Most seed growth curves are constructed by sampling seeds developing from flowers that were pollinated at roughly the same time, so they represent growth of an individual seed. These seeds can be identified by position in the inflorescence (e.g. wheat, rice, maize, etc.) or by marking fruits that are the same size (and therefore

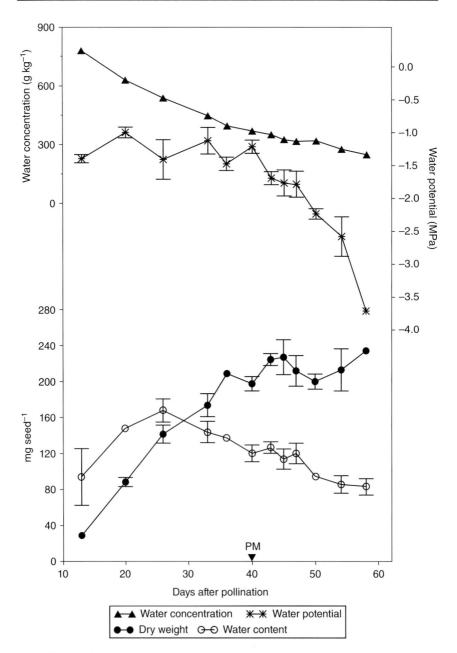

Fig. 2.4. Seed dry weight, water content and concentration, and embryo water potential of an individual F$_1$ maize (B73x Mo17) seed developing in the field. Bars represent ± 1 standard error of the mean. Standard errors of the mean for water concentration were smaller than the symbols. PM, physiological maturity. Adapted from Egli and TeKrony (1997).

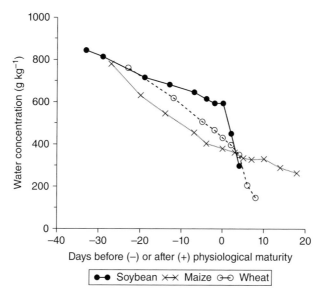

Fig. 2.5. Changes in water concentration during seed development (before physiological maturity) and during the desiccation phase (after physiological maturity) for soybean, maize and wheat seeds. Adapted from Egli and TeKrony (1997).

Table 2.1. Species variation in seed water concentration at physiological maturity.

	Caryopsis			Non-endospermic true seed		
Species	Water concentration (g kg⁻¹)	Source	Species	Water concentration (g kg⁻¹)	Source
Wheat	370–437	3, 4, 7	Soybean	550–600	7, 16, 21
Maize	337–377	2, 7, 13, 20	Bean	520–535	6, 22
Oat	450	8	Broad bean[3]	510–600	18
Barley	420–480	5, 12	Pea	550	17
Triticale	400	1	Chickpea	600	14
Pearl Millet[1]	350	9	White lupin[4]	600–650	14
Sunflower[2]	380–410	11, 15, 19			
Sorghum	320	10			

1. Bishnoi (1974). 2. Brooking (1990). 3. Calderini *et al.* (2000). 4. Clarke (1983). 5. Copeland and Crookston (1985). 6. Coste *et al.* (2001). 7. Egli and TeKrony (1997). 8. Frey *et al.* (1958). 9. Fussel and Dwarte (1980). 10. Gambin and Borras (2005). 11. Gesch and Johnson (2012). 12. Harlen and Pope (1923). 13. Hunter *et al.* (1991). 14. Jeuffroy and Ney (1997). 15. Kole and Gupta (1982). 16. Munier-Jolain *et al.* (1993). 17. Ney *et al.* (1993). 18. Pokojska and Gizelak (1996). 19. Rondanini *et al.* (2007). 20. Sala *et al.* (2007b). 21. Trawatha *et al.* (1993). 22. Van de Venter *et al.* (1996). [1]*Pennisetum glaucum*. [2]*Crysala* not a caryopsis. [3]*Vicia faba* L. var. minor. [4]*Lupinus albus*.

the same age) on a given day (e.g. soybean or other grain legumes). These curves provide a more precise description of seed growth than curves based on samples of all seeds on a plant or in a community (i.e. per unit area). All flowers on a plant or in an inflorescence are not pollinated at the same time, so the individual seeds start growing, enter and exit the linear phase and reach physiological maturity at different times, although the variation in the beginning of seed growth is usually larger than the end of seed growth (physiological maturity) (Hay and Kirby, 1991). A composite curve represents the summation of these seeds developing at different times and will not necessarily be the same as any individual seed. The length of the linear phase and the timing of physiological maturity of the composite curve could differ from curves based on individual seeds. Individual seed curves provide a more precise representation of seed growth characteristics and they are not influenced by changes in the number of seeds per plant during seed filling.

The difference between composite and individual seed curves depends upon the variation in time of pollination of the flowers on a plant or in an inflorescence. For example, there was 35 days or more between development of the first and last fruits on a soybean plant (Egli and Bruening, 2006a), although up to 84% of the fruits were initiated in less than half of the total period. Variation is usually less for crops with more compact fruiting structures, for example, the range was four to eight days in maize (Tollenaar and Daynard, 1978; Bassetti and Westgate, 1993a), four days in wheat (Evans *et al.*, 1972), and 11 to 12 days in oat (Rajala and Peltonen-Sainio, 2004). Differences between individual and composite seed curves could be relatively insignificant when there is little variation in flowering time or when changes in seed number during seed filling are small. For example, the total seed growth rate per unit area (g m^{-2} d^{-1}) is linear during most of the seed-filling period for most crops, mimicking the constant individual seed growth rate (mg seed^{-1} day^{-1}) during the linear phase, in spite of variation when individual seeds enter the linear phase. Estimates of the seed-fill duration will always be shorter for individual seeds than for a composite curve, although both estimates will probably capture genetic or environmental effects.

These general patterns of seed growth and development are followed by seeds of all crop species, although the time required for each phase varies within and among species and environments. Adams and Rinne (1980) also described a fourth phase, germination, which represents the establishment of the next generation. The focus of this book is on the role of the seed in the production of yield, so I will not cover germination; excellent coverage of germination is available from many sources, including, for example, Bewley *et al.* (2013, pp. 133–179) or Black *et al.* (2006).

Development of seed structures (Phase I)

Pollination and fertilization initiates phase I and is followed by a period of rapid cell division until all seed and fruit structures are present. Embryogenesis of many species has been described in detail (see Bewley *et al.* (2013, pp. 27–36) for excellent general coverage) and will not be repeated here. Detailed descriptions are available for soybean (Carlson and Lersten, 1987), maize (Kiesselbach, 1949), wheat (Huber and Grabe, 1987; Lersten, 1987), rice (Oparka and Gates, 1981), and many other crops.

Description of the early stages of development of a soybean fruit provides an excellent example of growth and development during phase I (Fig. 2.6). Sampling (day 1) started when the seeds were very small (< 5% of final mass), but the carpels (pod walls) had already reached their maximum length and width. Cell division in the cotyledons proceeded rapidly and cotyledon cells reached a maximum at approximately eight days, when the seed was still very small (< 15% of the maximum mass). Although there was no change in fruit size, the carpels continued to increase in dry weight during the 11-day sampling period. Carpel dry weight may increase significantly after phase I and, in some situations, decrease before PM suggesting redistribution of C and/or N to the seed (Fraser *et al.*, 1982a; Zeiher *et al.*, 1982). The soybean fruit in Fig. 2.6 reached the end of phase I on approximately day 8, when the pod was full size and cell division in the cotyledons had stopped. At this time, all of the pod and seed structures were formed and the

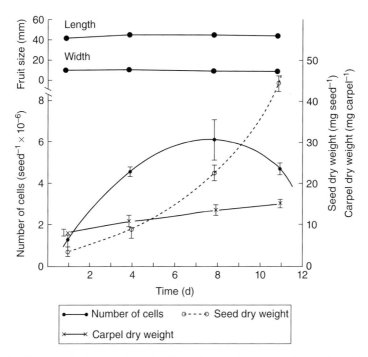

Fig. 2.6. Changes in the number of cells in the cotyledons and associated fruit and seed characteristics for an individual soybean seed. The date of the first sample was designated day 1 and the bars indicate ± 1 standard error of the mean. From Egli *et al.* (1981).

seed was ready to begin phase II, the rapid accumulation of storage reserves. Bils and Howell (1963) reported similar patterns for soybean. There are some reports of cell division continuing during the period of rapid accumulation of reserve materials (phase II) (e.g. Capitanio *et al.*, 1983 (maize); Guldan and Brun, 1985 (soybean); DeKhuijzen and Verkerke, 1986 (broad bean); Carceller and Aussenac, 1999 (wheat)). Whether there is a significant increase in the number of cells after phase I is hard to determine, partially because of the technical difficulties precluding accurate determination of cell number (note the apparent decline in cells after day 8 in Fig. 2.6). It seems likely, however, that the widely reported constant growth rate during Phase II is associated with a constant cell number, supporting the general consensus that cell division essentially stops at the end of phase I.

Patterns of development similar to those in Fig. 2.6 have been reported for wheat (Wardlaw, 1970; Gao *et al.*, 1993), maize (Reddy and Daynard, 1983; Jones *et al.*, 1996), sunflower (Lindstrom *et al.*, 2006) and rice (Zhang *et al.*, 2009) and other species. These general patterns of development are probably followed by all crop plants (Bewley *et al.*, 2013, pp. 85–92).

The linear phase of seed development (Phase II)

Dry matter accumulation and water relations

The linear phase of seed development begins at the end of phase I when cell division is essentially complete. Cell number is now fixed at its maximum and the rate of dry matter accumulation is constant with time (assuming constant environmental conditions), giving rise to the term 'linear phase'. The rate of seed respiration per seed is also constant during the linear phase of growth (Egli and Wardlaw, 1980) which, because of the constantly increasing seed mass, results in a steadily declining respiration rate per unit dry weight (Guldan and Brun, 1985). The seed growth rate will respond to changes in environmental conditions (e.g. temperature) or changes in the supply of assimilate to the seed during Phase II. These effects will be discussed at length in Chapter 3.

Since cell number is fixed during this phase, most of the increase in seed dry weight is the result of the accumulation of storage reserves which does not increase the seed's ability to accumulate dry matter (Harlan, 1920). Consequently, there is no reason to expect a constantly changing growth rate during this phase. In contrast, the growth of isolated seedlings, for example, follows an exponential growth curve because the new leaf tissue increases the photosynthetic activity per plant, resulting in an ever-increasing absolute growth rate. This relationship does not apply to seeds, because the metabolic machinery (cell number) is constant during the linear phase of seed growth. In spite of these well known growth characteristics, there have been suggestions that non-linear functions (e.g. various sigmoid polynomials or more complicated models) provide a better description of dry matter accumulation by seeds during phase II (Carr and Skene, 1961; Zahedi and Jenner, 2003). The normal sampling variation associated with estimating seed dry weight makes it difficult, if not impossible, to statistically distinguish between linear and non-linear models. Non-linear models, however, produce a constantly varying growth rate during seed filling which is not consistent with the mechanisms controlling seed growth discussed earlier. Consequently, there is no theoretical basis justifying non-linear functions, so the simple linear model of seed dry weight accumulation during phase II is widely accepted.

The increase in seed water content during phase II (Figs 2.2–2.4) reflects the movement of water into the cells to drive cell expansion, thus changes in seed water content and seed volume are closely associated. Cell number is constant during this phase of seed growth; consequently the large increase in seed size is entirely a result of increases in cell volume. Maximum seed water content represents maximum seed volume which occurs much earlier in phase II for cereals (Figs 2.3, 2.4) than for legumes (Fig. 2.2). There are reports of seeds reaching maximum volume after seed water content reaches a maximum (Sala *et al.*, 2007a; Gambin *et al.*, 2007), but no mechanism was put forward to explain an increase in volume without movement of water into the seed. The ability of the cells to increase in volume may regulate, in part, final seed size, suggesting that seed water relations may play a regulatory role in seed development (Walbot, 1978). This facet of seed development will be discussed in greater detail in later chapters.

Nutrient supply

Mature seeds are composed of many complex molecules, including storage proteins, lipids with a wide range in fatty acid composition and starch that is primarily made up of two polymers, amylose and amylopectin. Other polymeric sugars, including hemicelluloses and glucans, may accumulate in cereal and legume seeds (Bewley *et al.*, 2013, pp. 11–13) but are usually present in lesser amounts. Plants transport relatively simple compounds in the phloem (i.e. amino acids and sucrose), suggesting that these complex storage materials are synthesized in the seed from relatively simple precursors. Defining the site of storage reserve synthesis relates directly to a fundamental character of seed growth: is the seed simply a passive receptacle for assimilate or does the seed play an active role in regulating its own dry matter accumulation and its destiny? The seed, as we shall see, can synthesize complex molecules from simple raw materials and is anything but a passive receptacle for organic compounds supplied by the mother plant. The ability of the seed to partially regulate its own growth is the character that gives the seed a major role in the yield production process.

Sucrose is the primary source of carbon for seeds of crop plants. Numerous studies using ^{14}C and ingenious schemes for sampling fruit and seed tissues in several crop species have shown that sucrose is by far the most common sugar imported by the seed. This is entirely consistent with its role as the primary transport sugar in most plants.

Much of the carbon for seed growth comes directly from photosynthesis in leaves, but starch and other carbohydrates accumulated in vegetative plant parts of many crop species can be remobilized to become a source of carbon for seed growth. The contribution of remobilized carbon to yield is probably never very large. Data summarized by Evans (1993, pp. 254–258) suggests that the maximum contribution in non-stressed crops rarely exceeds 20%, although the contribution can increase when plants are exposed to drought stress during seed filling (Sadras and Conner, 1991).

Many seed or fruit structures are photosynthetically active (Sambo *et al.*, 1977; Caley *et al.*, 1990; Watson and Duffus, 1991; Whitfield, 1992; Eastmond *et al.*, 1996; Furbank *et al.*, 2004) and carbon fixed by these structures also contributes to seed growth. Seeds often develop in low-light environments, created by shade from leaves, carpels, siliques or glumes, greatly limiting photosynthetic contributions (Eastmond *et al.*, 1996). Many reproductive structures show a net efflux of CO_2 in the light (Sambo *et al.*, 1977) so their primary contribution is through a partial refixing of respiratory CO_2. Bort *et al.* (1996) estimated that 55 to 75% of the respired CO_2 was refixed in ears of barley and wheat, while CO_2 refixed by soybean pods accounted for 4 to 16% of the total fruit import (Sambo *et al.*, 1977; Layzell and LaRue, 1982). Caley *et al.* (1990) summarizing results from the literature found that 10 to 76% of seed dry weight came from ear photosynthesis in barley and wheat. Obviously the photosynthetic contribution of the reproductive structures depends on whether they are in a high light environment above the canopy (wheat, barley and rice, for example) or the low light environment in or below the plant canopy (e.g. soybean or maize).

Seeds cannot utilize inorganic forms of N; they require an organic source and a variety of amino acids and ureides are delivered to the fruit and seeds via the phloem. Although ureides are found in the phloem of fruit tissues of some species, they are apparently converted to other forms of organic N in the maternal tissues and do not reach the embryo. Only a few of the many amino acids supplied to the seed provide most of the seed's N supply. For example, in soybean only two amino acids (asparagine and glutamine) of the 17 identified in seed coat exudates accounted for 75% of the N supplied to the embryo (Rainbird *et al.*, 1984). Three amino acids (alanine, asparagine and glutamine) accounted for 63% of the total of 22 amino acids identified in seed coat exudates of broad bean (Wolswinkel and de Ruiter, 1985). Alanine and glutamine accounted for 40% of the total N in seed coat exudates from nodulated pea, (Rochat and Boutin, 1991). Approximately 20 amino acids were detected in the phloem sap of rice, but four (serine, asparagine, glutamine and glycine) predominated (Fukumorita and Chino, 1982), although, there were some changes in relative quantities during seed development. Glutamic acid and aspartic acid predominated in the phloem sap of wheat (Hayashi and Chino, 1986).

There may be species variability in the amino acids supplied to the developing seed, but all species seem to get much of their seed N from just a small group of amino acids. The profile of amino acids supplied to the seed does not have to quantitatively match the amino acid profile of the storage proteins because the seed has the ability to synthesize amino acids in the amounts and proportions needed.

The triacylglycerols that make up most plant oils are fatty acids esterified to the hydroxyl groups of glycerol. Seeds synthesize the fatty acids and glycerol from sucrose that is supplied by the mother plant. In fact, all seed oils and complex carbohydrates – starch, cellulose, lignin, complex sugars, etc. – must be synthesized from sucrose imported from the mother plant.

The ability of seeds to grow relatively normally *in vitro* in simple nutrient solutions clearly demonstrates that they can synthesize storage reserves from a few simple precursors. For example, Thompson *et al.* (1977) demonstrated near normal rates of dry matter and protein accumulation by soybean cotyledons in nutrient media containing sucrose, one amino acid, and various mineral nutrients. Similar results have been reported by Hayati *et al.* (1996) (soybean), Barlow *et al.* (1983) (wheat), and Cully *et al.* (1984) (maize).

Much is known of the pathways for synthesis of seed storage compounds. A detailed treatment of seed biochemistry is beyond the scope of this book, but excellent coverage of this subject can be found in Bewley *et al.* (2013, pp. 96–129).

Seeds have the ability to synthesize complex molecules in relatively precise amounts and proportions so they are not simply passive receptacles for the C and N assimilates and mineral nutrients provided by the mother plant. Many facets of seed growth are regulated by the seed, not by the supply of raw materials to the seed, but the seed cannot grow without a supply of raw materials from the mother plant. The regulation of some aspects of its own destiny provides an important role for the seed in the production of yield. If the seed simply served as a container that passively accumulated raw materials supplied by the plant, there would be no need to consider seed growth as a part of the yield production process and no need for this book.

Seed composition

The economic value of seeds is derived from the oil, protein and starch synthesized during seed development. Synthesis of the storage materials by the seed from sucrose and a few amino acids suggests that the proportions of oil, protein and carbohydrate as well as the composition of these storage materials are, at least partially, regulated by the seed, not by the supply of C and N from the mother plant. Control by the seed is evident when genetic differences in protein concentration in soybean and maize seeds were maintained during *in vitro* growth over a range of N concentrations (Wyss *et al.*, 1991; Hayati *et al.*, 1996). Since seeds cannot grow without C and N from the mother plant, it is not surprising that the supply can also influence seed composition. *In vitro* culture studies have clearly shown that protein, oil and carbohydrate concentration responds to variation in N supply (Wyss *et al.*, 1991; Hayati *et al.*, 1996). Carbon availability during seed filling can also affect seed composition (Echarte *et al.*, 2012).

The regulation of seed composition by the supply of raw materials or by the seed can be better understood by considering the two sources of variation in seed composition – environmental and genetic. Environmental conditions can modulate the supply of assimilate and N to the seed, which could, in turn, affect seed composition, whereas genetic differences in seed composition are probably regulated by the seed (Wyss *et al.*, 1991; Hayati *et al.*, 1996). Effects of the environment on the supply of assimilate, N and seed composition are probably minimized by changes in seeds per plant or per unit area in response to variation in canopy photosynthesis and the supply of assimilate (see Chapter 4), which would maintain a relatively constant supply per seed. Maintaining a constant supply of assimilate per seed may contribute to the relative stability of seed composition among environments. The environment can also affect seed composition through direct effects on seed metabolism. The effect of temperature (e.g. oil concentration in soybean seeds increases as temperatures increase, while seed protein concentrations decrease, Wilson, 2004) is likely to be a direct effect on metabolism in the seed, not on the supply of assimilate to the seed.

Seed composition at maturity represents the integrated effect of synthesis activities throughout seed development, but the rate of synthesis of seed components is not always constant during seed filling. Protein concentration in soybean and sunflower seed is nearly constant during seed development, indicating a constant rate of synthesis, but oil concentration is initially low and then it increases, reaching its maximum level as the seed approaches physiological maturity (Yazdi-Samadi et al., 1977; Ruiz and Maddonni, 2006). Starch reaches a maximum level early in seed development in soybean (Egli and Bruening, 2001) and other high-oil species before declining to nearly zero at physiological maturity. Apparently, starch accumulates when there is little oil synthesis early in seed development and declines as the rate of oil synthesis increases at later stages (Bewley et al., 2013, p. 98). In contrast, starch concentration in developing wheat seeds (a low-oil seed) did not change during development (Jenner and Rathjen, 1977). Variation in the rate of synthesis of various carbohydrates, storage proteins and fatty acids during development has been reported for several species (Wilson, 1987).

Variation in seed composition during development may contribute to environmental effects on mature seed composition. Premature cessation of seed growth would produce a mature seed that reflected the composition of the seed when growth stopped, not the composition at normal maturity (Wilson, 1987). For example, drought stress during seed filling often causes premature seed maturation in soybean (de Souza et al., 1997) and other crops. Seed oil concentration increases during seed development in soybean and drought stress often decreases it (Rotundo and Westgate, 2010), an effect that is consistent with composition determined by the stage of development when seed growth stops. The duration of seed fill of individual seeds of some species may be related to the time of pollination of individual flowers with seeds from late developing flowers often having shorter developmental periods. If the shorter developmental period represents premature cessation of growth, these differences in duration could affect seed composition, creating variation among seeds on the plant.

Mature seed composition is controlled by the seed's genetic makeup and the environment (assimilate supply and temperature) in which it develops. Control therefore resides in both the mother plant and in the developing seed. Genetic variation, regulated by the seed, is probably more important than environmental effects and it has long been exploited by plant breeders and now by biotechnologists to improve the usefulness of seeds.

Hormones

Seeds are rich sources of plant hormones. Auxins, gibberellins, cytokinins, brass-inosteroids, ethylene and abscisic acid are all found in seeds (Bewley *et al.*, 2013, pp. 36–46); in fact, seeds were the first higher-plant source for most of these hormones. Hormones play important regulatory roles in seed growth, including, among other possible roles, involvement in growth and development of the seed and accumulation of storage reserves and their use during germination and early seedling growth (Bewley *et al.*, 2013, pp. 36–46; Jones and Setter, 2000). The potential role of hormones in seed growth, development and germination has been reviewed elsewhere (Bewley *et al.*, 2013, pp. 36–46) and the reader should consult this review for more details.

The end of seed growth – physiological maturity (Phase III)

Physiological maturity is defined as the occurrence of maximum seed dry weight and represents the end of dry weight accumulation and the seed-filling period. This definition of physiological maturity was probably first used by Shaw and Loomis (1950) in their research with maize. Others have referred to this growth stage as relative maturity (Aldrich, 1943), morphological maturity (Anderson, 1955) and, more recently, mass maturity (Ellis and Pieta-Filho, 1992). Physiological maturity has been widely adopted as an important growth stage and used by researchers and producers, because it represents the end of active plant growth and the production of yield.

The seed no longer has a functional connection to the vascular system of the mother plant at physiological maturity and assimilate no longer moves into the seed. The ^{14}C recovered from seeds after exposing the leaves to $^{14}CO_2$ decreased to very low levels when seeds of sorghum (Eastin *et al.*, 1973), soybean (TeKrony *et al.*, 1979) and maize (Hunter *et al.*, 1991) reached physiological maturity. Oat showed a similar pattern when the cut end of the panicle was allowed to take up ^{14}C-sucrose (Lee *et al.*, 1979). These results support the assertion that the seed is 'isolated' from the mother plant at physiological maturity and is essentially in storage on the plant.

Harvest maturity, when the seed has dried to a harvestable moisture level, occurs after physiological maturity. Seed moisture concentrations are relatively high at physiological maturity (Table 2.1), so the seed must dry before it can be harvested. Identifying harvest maturity is of little value in studies of yield physiology because the production of yield ends at physiological maturity. Plant and environmental factors that affect yield can only do so before physiological maturity. For example, including precipitation that occurs between physiological and harvest maturity in any evaluation of the relationship between water availability and yield only complicates the analysis, because precipitation falling after physiological maturity cannot affect yield. Yield, to the commercial producer, is harvested yield, which can be reduced in amount or quality by weather damage, disease or other problems occurring between physiological maturity and harvest. These losses are an important part of commercial production of any crop, but they are completely separate from the physiological processes that produced yield. The production of yield is complete at physiological maturity.

Seed water concentration at physiological maturity varies among crop species (Table 2.1) and there may also be some variation among cultivars within a species. Species variation seems to be associated with seed structure (Rondanini *et al.*, 2007) with species producing true, non-endospermic seeds having higher water concentrations at physiological maturity (510–650 g kg^{-1} across multiple species and sources) than those producing a caryopsis (320–480 g kg^{-1}). Sunflower produces a cypsela, but the water concentration at physiological maturity was similar to the cereals. Egli (1998, p. 30), however, theorized earlier that seed composition was the key determinant of water concentration at physiological maturity. Most of the low water concentrations occur in high starch, low oil and protein seeds, i.e. species producing a caryopsis, with the exception of sunflower with its high oil and protein levels. The high water concentrations are found in true non-endospermic seeds with high protein levels. Structure and composition effects in Table 2.1 are completely confounded; perhaps data representing species with more variation in composition will help clarify this issue.

Experiments comparing cultivars within a species suggest that there may also be significant cultivar differences in the water concentration at physiological maturity (Hallauer and Russell, 1962; Rench and Shaw, 1971; Daynard, 1972; Hunter *et al.*, 1991). Stress that caused premature seed maturation resulted in higher water concentrations at physiological maturity in some species (e.g. maize (Sala *et al.*, 2007b) and sunflower (Rondanini *et al.*, 2007)) but not in others (e.g. wheat) (see Rondanini *et al.*, 2007 for a summary). Increasing source–sink ratios lowered seed water concentration at physiological maturity in sorghum (Gambin and Borras, 2007), but not in maize (Sala *et al.*, 2007b). These differential effects of source–sink modifications support the contention (discussed in Chapter 3) that seed growth is controlled by both the seed and the supply of assimilate to the seed (Sala *et al.*, 2007a).

Accurate estimates of seed water concentration at physiological maturity require accurate estimates of when it occurs. Since seed water concentration often changes rapidly after physiological maturity, estimates of physiological maturity that are too late can result in seed water concentrations that are much too low. The method used to measure water concentration (Grabe, 1989) also contributes to the variation. Some of variation in seed water concentration at physiological maturity within and among species in Table 2.1 could be a result of these inaccuracies. Precise data and frequent samplings are needed to accurately estimate seed water status at physiological maturity.

The water potential of embryos or axes at physiological maturity was relatively constant for wheat (–1.7 MPa), maize (–1.6 MPa) and soybean (–1.5 MPa) (Egli and TeKrony, 1997), suggesting that the water status of important seed tissues at critical growth stages may be independent of seed type, composition and species. Water potential is probably a much better indicator of tissue water status than water concentration, so much of the variation in Table 2.1 may disappear if water potential replaced water concentration. It is not known whether the reported cultivar differences or stress and source–sink effects on water concentration at physiological maturity are associated with differences in embryo or axis water potential.

Seeds usually lose water relatively rapidly after physiological maturity because the seed is no longer attached to the vascular system of the plant and no longer receives water to replace that lost to the environment. Moisture loss after physiological maturity is determined by environmental conditions and crop species, and varies from very rapid (e.g. soybean, wheat) to relatively slow (e.g. maize) (Fig. 2.5). Plant and seed structures in maize (Fig 2.5) and sorghum (Gambin and Borras, 2005) restrict water movement from the seed to the atmosphere, slowing the decline in water concentration to a rate similar to that before physiological maturity. In other crops, e.g. soybean and wheat, the restrictions to water movement are less and the decline in water concentration after physiological maturity is much faster. Seed moisture may increase again in extremely wet environments during the drying phase and it can reach a level that triggers premature germination.

The relatively high moisture concentration at physiological maturity suggests that metabolic activity may not stop when assimilate is no longer imported from the mother plant. Howell *et al.* (1959), Ohmura and Howell (1962) and TeKrony *et al.* (1979) found that soybean seed respiration declined as seeds approached physiological maturity, but it was not zero at physiological maturity, indicating that the seeds were still metabolically active even though they no longer received assimilate from the mother plant. Similar results were reported for pea (Kolloffel and Matthews, 1983) and pearl millet (Fussell and Dwarte, 1980).

Extensive seed respiration between physiological and harvest maturity could cause reductions in seed dry weight and yield. If these loses are significant, harvesting at high seed moisture levels or the use of desiccants to encourage rapid drying could increase harvestable yield. Ashley and Counce (1993) reported significant losses in dry weight of cereal grains after physiological maturity and the losses were greater in years with high rainfall. Most published seed growth curves show little decline in seed dry weight after physiological maturity (e.g. Figs 2.2–2.4, Gambin and Borras, 2005, sorghum; Tang *et al.*, 2009, rice), probably reflecting the usual rapid drying and cessation of physiological activity. The sampling variation associated with estimates of seed dry weight make it difficult to detect small decreases if they occurred, but there seems to be little compelling evidence of significant loss of weight after physiological maturity.

The occurrence of physiological maturity also has important implications for seeds used to regenerate the crop. These seeds must have the capacity to germinate and produce a healthy seeding when planted in the field. Many researchers believe that maximum seed germination and vigour occur when the seed reaches its maximum weight, i.e. at physiological maturity (Harrington, 1972), although this concept has been challenged by Ellis and co-workers (Ellis and Pieta-Filho, 1992). They suggested that maximum seed vigour occurs after physiological maturity and they proposed the term 'mass maturity' to separate the occurrence of maximum seed dry weight from maximum seed vigour thereby assigning physiological maturity to maximum seed vigour. Seeds at physiological maturity must be dried before vigour levels can be determined and the results of TeKrony and Egli (1997) suggest that the relationship between the occurrence of maximum seed weight (physiological maturity) and maximum seed vigour may depend upon how the seeds are harvested and dried before vigour is determined. There is no straight-forward answer to the question of when maximum seed vigour occurs, but there seems to be no compelling reason to abandon the use of physiological maturity to refer to maximum seed dry weight. Physiological maturity is still widely used to refer to its original meaning of maximum dry weight, and I will follow this convention in this book.

Determining physiological maturity

Physiological maturity is an important growth stage, because it represents the end of the yield production process to the crop physiologist, and, to the agronomist and crop management specialist, the time when yield will no longer be affected by weather, diseases, insects or crop management decisions (ignoring harvest losses and deterioration in quality). Accurate determinations of the occurrence of physiological maturity are very useful and the most useful techniques are those that are easy, non-destructive, not subjective, require no specialized equipment and can be quickly applied in the field.

Physiological maturity can be determined for an individual seed, a single plant, or a plant community. Although physiological maturity can usually be determined fairly precisely for an individual seed, determination for a plant or community (field) is more difficult because all of the seeds on a plant or in a plant community will not reach physiological maturity at the same time. Absolute physiological maturity of a plant occurs when all of the seeds have reached physiological maturity; however, absolute physiological maturity is not needed for most practical applications. The total seed weight of a plant at absolute physiological maturity would not differ much from a plant with, for example, 60% of the seeds at physiological maturity. The dry weight of seeds approaches physiological maturity asymptotically (Figs 2.2–2.4), so there is little weight gain in the last few days before physiological maturity. The failure to have all of the seeds at physiological maturity is of little practical consequence in terms of yield (TeKrony et al., 1981). Waiting for the last few seeds to mature could reduce the effectiveness of management practices scheduled at physiological maturity. The date of physiological maturity can vary widely within a field and this must be taken into consideration when scheduling activities dependent upon the occurrence of physiological maturity.

In theory, measurement of seed dry weight changes with time provides a direct estimate of physiological maturity (i.e. when dry weight reaches its maximum). However, seed dry weight approaches its maximum asymotopically (Fig. 2.7) which, when combined with the normal sampling variation associated with measurements of seed dry weight (Fig. 2.4), makes it very difficult to estimate accurately when maximum dry weight occurs. Estimates based on individual data points (e.g. time when there is no significant difference from the previous sample (Rondanini et al., 2007)) will be affected by the sampling interval – daily or every-other-day

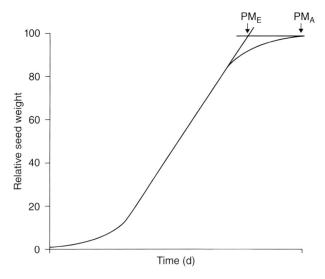

Fig. 2.7. Diagrammatic representation of the determination of physiological maturity from dry weight measurements using the 'broken stick' method described by Crookston and Hill (1978). PM_A, actual date of physiological maturity, i.e. maximum seed dry weight; PM_E, estimated date of physiological maturity.

measurements are needed as the seed approaches its maximum weight, to determine precisely when physiological maturity occurs.

A number of techniques have been used to estimate the occurrence of maximum dry weight from seed dry weight data. Crookston and Hill (1978) and TeKrony and Egli (1997) estimated the maximum seed dry weight with the average of all samples that did not differ statistically from the most mature sample. The remaining samples were used to estimate a linear equation representing the accumulation of weight during seed development, and the time when the linear curve intercepted the maximum seed weight was taken as physiological maturity (PM_E in Fig. 2.7). Estimating the point of intersection of the two straight lines with an iterative regression approach has become a popular technique (Pieta-Filho and Ellis, 1991; Rondanini *et al.*, 2007; Borras *et al.*, 2003). Physiological maturity estimated by this linear-plateau or 'broken stick' procedure occurs before the actual occurrence of maximum seed dry weight, as illustrated diagrammatically in Fig. 2.7. One advantage of this technique is that it is not subjective; however, it requires destructive sampling to construct the entire seed growth curve.

Physiological maturity can also be estimated from regression equations representing a complete seed growth curve. Hanft and Wych (1982) fit a cubic polynomial equation to wheat seed growth curves and calculated the time of occurrence of maximum seed weight, while Smith and Donnelly (1991) used a splined regression analysis involving two third-order polynomials. In both cases, their estimates of physiological maturity seem to be after the actual date because the seed moisture concentrations at their estimated date were lower than other published estimates.

In fact, Hanft and Wych's (1982) estimates of 95% of maximum seed weight may represent a better estimate of physiological maturity. This technique also requires construction of a complete seed growth curve and its accuracy depends upon how well the regression model represents the true seed growth curve.

Estimating physiological maturity from seed dry weight curves is time consuming, requires destructive sampling and does not produce the real-time estimate needed for management decisions. The time-consuming nature of dry weight approaches fuelled the search for indirect indicators of physiological maturity that were quick, non-destructive, accurate and easy to use. Non-subjective visual indicators meet these requirements.

Seed moisture concentration at physiological maturity is relatively stable within a species and can be used as an indirect indicator of physiological maturity. Seed water concentration works better in some species (e.g. soybean) where the rate of decline in water concentration increases substantially after physiological maturity than in species (e.g. maize) where there is little change after physiological maturity. This method suffers from the same disadvantages as dry weight, requiring frequent destructive sampling and, in addition, it requires accurate measurements of seed water concentration. Variation in water concentrations at physiological maturity across cultivars and possibly environments (Hunter *et al.*, 1991) also complicates use of this technique. One advantage of seed water concentration is that it is usually less variable than seed dry weight. Estimating physiological maturity from seed water concentration is indirect, but it does not meet the criteria of being quick, easy to use and non-destructive; indirect indicators based on plant or fruit/seed characteristics meet this criterion.

The characteristics of vegetative plant senescence (leaf yellowing or leaf abscission) have also been used as indicators of physiological maturity, but this approach usually requires subjective evaluations of the degree of leaf yellowing or leaf abscission. The completion of senescence is generally associated with the end of seed filling (i.e. seeds cannot grow without a source of assimilate), but there can be significant environmental and cultivar variation in this relationship. Seeds will mature when assimilate is still available (see Chapter 3) (Banziger *et al.*, 1994), so complete senescence is not an absolute requirement for physiological maturity, further confounding the relationship between leaf senescence and physiological maturity. Descriptions of leaf senescence are usually not reliable indicators of physiological maturity (Housley *et al.*, 1982).

SOYBEAN. Research with $^{14}CO_2$ labelling techniques demonstrated that the amount of ^{14}C recovered from seeds declined as the colour of the seeds and pods changed from green to yellow (TeKrony *et al.*, 1979). Very little ^{14}C was recovered from yellow seeds, even if they were found in pods that were not completely yellow. Thus, physiological maturity in a soybean seed occurs when the seed turns yellow. Pod walls and seeds usually change from green to yellow at roughly the same time, but it is possible to find completely yellow seeds in pods that are not completely yellow. Colour change of seeds probably provides a better estimate of physiological maturity than pods. Crookston and Hill (1978) reached a similar conclusion, although they associated physiological maturity with the loss of green colour from seeds.

In field soybean communities that were at physiological maturity, estimated from seed dry weight data, only 26% of the seeds, averaged across six cultivars and several planting dates, were yellow, but 70% were either green/yellow (i.e. > 50% green) or yellow/green (> 50% yellow) and were probably in the final lag phase of growth (Fig 2.7) (TeKrony *et al.*, 1981). It was impossible to detect significant differences in yield between plots harvested at growth stage R7 (beginning maturity – one normal pod on the main stem has reached its mature pod colour, Fehr and Caviness, 1977) and at full maturity (95% brown pods) in several field experiments even though 45–50% of the seeds were not completely yellow at R7. The daily increments in seed dry weight are so small as the seed approaches physiological maturity that yield from harvests slightly before absolute physiological maturity are not measurably different from yield at full maturity. Growth stage R7, as defined by Fehr and Caviness (1977), is now generally accepted as an easily determined indicator of physiological maturity of individual plants, even though all of the seeds on the plant have not reached physiological maturity at this stage. Growth stage R7 of a community is usually defined as occurring when 50% or more of the plants are at or beyond growth stage R7 (Fehr and Caviness, 1977). Harvest maturity (the first time the seeds dried to 140 g kg^{-1} water concentration) occurred from 9 to 24 days after growth stage R7, depending on environmental conditions (TeKrony *et al.*, 1981).

MAIZE. The appearance of a layer of brown crushed cells – the black layer – in the placental-chalazal region of the seed that closes the hilar region provides an estimate of physiological maturity in maize seed. This layer was described by Johann (1935) and Kiesselbach and Walker (1952), while Daynard and Duncan (1969) evaluated its usefulness as an indicator of physiological maturity. Labelling studies with ^{14}C and changes in dry weight also indicated that formation of the black layer signalled the end of assimilate movement into the seed (Hunter *et al.*, 1991). The black layer develops gradually and it is easy to identify seeds that have reached complete black layer, but it is harder to consistently identify the intermediate stages. Hunter *et al.* (1991) modified the five intermediate stages described by Rench and Shaw (1971), and found that maximum seed weight occurred at black layer stage four, in which a thin, dark-brown band (usually ≤ 1 mm thick) reaches across the entire base of the seed between the junction of the basal endosperm and embryo tissue with the pedicel-placental tissues.

The milk line, the line on the abgerminal face of the maize seed, dividing solid from liquid endosperm, is also a useful indicator of physiological maturity. The milk line is easily observable and develops as the endosperm starts to solidify. There is no milk line in an immature seed where the entire endosperm is liquid, but as the endosperm solidifies, the milk line moves down the kernel from the top until all of the endosperm is solid and there is again no milk line present. Hunter *et al.* (1991) found that the milk line and black layer developed in parallel and milk line stage four (75% of the seed's length contains solidified endosperm; milk line is present near the seed's base) represented physiological maturity. Afuakwa and Crookston (1984) reported that 95% of maximum seed dry weight had accumulated when the milk line reached the mid-point of the seed. Both the milk line and black layer provide practical and useful visual indicators of physiological maturity of maize seeds.

WHEAT.   Hanft and Wych (1982) related 13 visual characteristics of plants and seeds of eight hard red spring wheat cultivars to physiological maturity estimated by fitting polynomial regression equations to seed dry weight curves. Their estimate of 95% maximum seed weight, which probably represents a better estimate of physiological maturity from their regression analysis than maximum seed weight, was closely associated with a loss of green colour in the flag leaf or the first appearance of a dark pigment strand beneath the embryo in the seed. Sofield *et al.* (1977) reported that maximum seed weight was associated with deposition of lipids in the pigment strand. Smith and Donnelly (1991) found that the pigment strand was difficult to observe in their material and concluded that loss of green colour from most portions of the spike provided the best indicator of physiological maturity. Housley *et al.* (1982) associated maximum seed dry weight with the onset of the development of red colour in the seeds (10% of the maximum colour of the mature seeds) which occurred before the darkening of the pigment strand beneath the embryo. Among the variety of plant and seed characters evaluated in wheat, the change in colour of the seed, seed structures or the spike were generally associated with physiological maturity.

SORGHUM.   Eastin *et al.* (1973) demonstrated with $^{14}CO_2$ labelling that physiological maturity was associated with the appearance of a dark closing layer in the placental area near the point of attachment of the sorghum seed to the mother plant.

BARLEY.   Harlan and Pope (1923) associated development of black colour in the pericarp with maximum seed weight in a 'naked' (seed free from the glumes at maturity) barley cultivar. The loss of green colour from glumes and peduncle was closely associated with 95% maximum dry seed weight for several cultivars grown in the field for two years (Copeland and Crookston, 1985). Their estimates of 95% maximum seed weight came from cubic polynomials fit to seed dry weight curves.

OAT.   Physiological maturity, based on maximum seed dry weight and when the movement of water-soluble dye and ^{14}C-sucrose into isolated panicles stopped, occurred when 75% of the glumes were yellow (Lee *et al.*, 1979).

OTHER CROPS.   A dark closing layer was found in pearl millet seeds (Fussell and Dwarte, 1980) at physiological maturity. Physiological maturity in sunflower was associated with floret abscission (Browne, 1978) or the back of the heads and involucral bracts turning yellow (Robinson, 1983). These changes in sunflower were associated with growth stage R9 (Schneiter and Miller, 1981).

Visual indicators of physiological maturity have been developed for many crops and they are often based on changes in seed colour or seed characteristics. These indicators are useful in both research and crop management, because they provide a quick and easy determination of when seed growth has stopped and when yield will not be affected by the environment or manipulating the plant. The seed is now in storage on the plant as it enters its final phase of development, drying to a moisture level suitable for harvesting and storage.

The accuracy required of an estimate of physiological maturity depends on the use of the estimate. Physiological studies of individual seeds frequently require estimates within one or two days of the actual event. For example, investigations of changes in seed water status during development and maturation can be misleading if the estimate of maturity is off by only a few days, because seed water status frequently changes rapidly after physiological maturity. Most methods will indicate physiological maturity of plants or plant communities before all seeds reach their maximum weight. because of the variation in the timing of development among seeds on a plant or among plants in a community. This variation is even greater on a large field scale, where plants in different locations in the field may reach physiological maturity at quite different times. Fortunately, using physiological maturity for field scale crop management decisions does not require highly accurate estimates of its occurrence. Management actions (e.g. early harvest or herbicide applications to kill late-season weeds) affecting yield can usually be applied slightly before or after all seeds or plants have reached physiological maturity without serious consequences.

Summary

The general patterns of growth and development are the same for seeds of all common crop species, regardless of their structure, composition or size. Consequently, we can treat these seeds as a common group to investigate the role of the individual seed in the production of yield.

The developing seed, a mixture of maternal and embryonic tissues, is dependent upon the mother plant for the nutrients that sustain its growth. The seed, however, is not just a passive storage container that accumulates the nutrients supplied by the mother plant. Instead, the seed synthesizes its storage reserves from sucrose and amino acids arriving in the phloem. Photosynthesis in vegetative plant parts is the primary production process behind the supply of nutrients to the seed, but it is only part of the yield production process in grain crops. The synthesis and accumulation of storage reserves in the seed are equally important and the seed plays a central role in this part of yield production process. It is this central theme that we will investigate in this book.

The final mature seed dry weight can be described as a function of the rate of dry weight accumulation (mg seed^{-1} day^{-1}) and the length (days) of the dry weight accumulation period. We will use these two parameters, seed growth rate and the duration of seed growth, to study the factors affecting seed growth and their relationship to yield in the following chapters.

Seed Growth Rate and Seed-fill Duration: Variation and Regulation

3

The growth of the seed that is harvested for economic yield in grain crops has two components – a rate component and a time component. I defined these in Chapter 2 as the seed growth rate (SGR) and the seed-fill duration (SFD). Variation in final seed size (weight per seed) occurs because seeds grow rapidly or slowly for longer or shorter periods.

We cannot understand the central role of the seed in the yield production process without a thorough evaluation of genetic and environmental variation in SGR and SFD, and the plant processes responsible for this variation. This evaluation will prepare us for Chapters 4 and 5, where we will consider the role of the seed in the production of yield.

The SGR can be measured on a community basis (i.e. total seed growth rate (TSGR), g m^{-2} day^{-1}), where it represents the average of all seeds in the community, or it can be determined for an individual seed (i.e. mg seed^{-1} d^{-1}). Total seed growth rate is more complex than individual SGR, because estimates at the community level include potential effects of plant characteristics and productivity of the environment, as well as the characteristics of the individual seed. Estimates at the individual seed level are devoid of most of these influences and reflect only the basic characteristics of the seed. For example, the TSGR is affected by species (C4 species often greater than C3) and the productivity of the environment, but these differences are usually a result of variation in seeds per unit area and are completely independent of the characteristics of the individual seed.

Seed-fill duration can also be estimated at the community or individual seed level. Estimates at the community level can be influenced by species variation in phenological development, and the uniformity of flowering, which is often related to how the seeds are borne on the plant (fruits at individual nodes, seeds in a panicle or in a single compact ear at the top or near the middle of the stem) and therefore differ from estimates at the individual seed level. The relatively long-flowering and pod-set period in soybean, for example, may result in a longer community SFD than estimates at the individual seed level. Species with a shorter flowering period may differ from species with a longer flowering period, even though the SFD of individual seeds is the same.

The difference between estimates of TSGR and individual SGR are probably greater than community vs individual seed estimates of SFD. Evaluating SGR and SFD on a single seed basis eliminates many of these confounding issues and provides a much clearer depiction of the characteristics of seed growth. Such a basic understanding is needed to understand the role of the seed in the production of yield, so I will focus on seed growth at the fundamental level of the individual seed.

The rate of accumulation of dry matter by the individual seed increases to a maximum during the early stages of seed growth and then slows to zero as the seed reaches its maximum weight at physiological maturity (Figs 2.2–2.4). Seed growth rate is commonly taken to represent the accumulation of dry matter with time during the linear phase of growth (phase II, as described in Chapter 2) (i.e. the maximum growth rate) and the accumulation during the lag phases at the beginning and end of seed growth are ignored. Ignoring the lag phases does not create a serious problem because most of the seed dry weight accumulates during the linear phase. The SGR is usually estimated with linear regression using seed dry weights collected at regular intervals during the linear phase of seed development. Since the change in seed dry weight with time is known to be linear, some researchers have estimated SGR from only two samples during the linear phase.

Polynomial, sigmoid and logistic functions have also been used to evaluate seed growth characteristics (e.g. Darrock and Baker, 1995). These functions require more data to adequately describe the entire growth curve and the estimates of SGRs are incorrect. The derivatives of these complex functions produce estimates of SGR that change throughout the seed-filling period (see, for example, Tang *et al.*, 2009), a misrepresentation of the true situation as discussed in Chapter 2. To argue that SGR increases to a maximum at the mid-point of the growth curve and then declines steadily until PM, with no known mechanism to produce such a pattern, represents the triumph of statistical modelling over physiological principles. Seed growth rate in this book will always refer to the growth rate of an individual seed estimated with linear regression during phase II (the linear phase) of seed growth.

The duration of seed growth is harder to determine, because it is difficult to accurately estimate when the seed starts and stops accumulating dry matter. On a community basis, anthesis or a whole plant growth stage (e.g. growth stage R5 in soybean) can be used to estimate the beginning of seed growth and physiological maturity the end, which, with indirect indicators of physiological maturity, provides a non-destructive estimate of SFD. This technique will produce community estimates of SFD that can vary among species, depending upon their fruiting characteristics. Anthesis to physiological maturity can be used for cereals, but a growth stage that defines the beginning of seed filling is needed for crops with long flowering periods. Maize often has a longer period from silking to physiological maturity than other cereals (e.g. wheat) because the time from silking to the beginning of seed growth is longer (Egli, 2004). These non-destructive estimates are useful for comparisons within a species, but they cannot be used to make valid comparisons among species. It is difficult, however, to apply growth stage techniques to individual seeds.

The effective filling period (EFP) is frequently used as a relative estimate of the length of the seed-filling period (Fig. 3.1). The EFP was first described by Hatfield and Ragland (1967) at the University of Kentucky on a per plant basis and by W.G. Duncan, T.B. Daynard and J.W. Tanner of the Universities of Kentucky and Guelph (Daynard *et al.*, 1971) on a community basis, as the grain yield (kg plant^{-1} or ha^{-1}) divided by the total rate of dry matter accumulation by the seeds (kg plant^{-1} day^{-1} or ha^{-1} day^{-1}) during the linear phase of seed growth. The EFP can also be calculated on an individual seed basis by dividing the mature weight per seed by the SGR. Either way, the EFP estimates the time it would take to produce yield or a mature seed if the seed(s) (individual or total per area) always grew at the linear rate (Fig. 3.1). This method avoids the problem of accurately estimating the beginning and end of seed growth and is easy to use in studies involving measurements of SGR (TSGR) because only an additional determination of final size (yield) is needed to complete the calculation. The EFP is independent of species differences in phenological development, so it is the best method for species comparisons of SFD. The EFP is a mathematical construct that provides an estimate of SFD; it is not a seed growth stage, although it is often used this way in the contemporary literature.

Statistical models of complete growth curves can also be used to estimate SFD by calculating, for example, the time from 5 to 95% or 10 to 90% of maximum seed weight (see Johnson and Tanner, 1972 for examples). The quality of the estimates depends upon how well the statistical function describes the seed

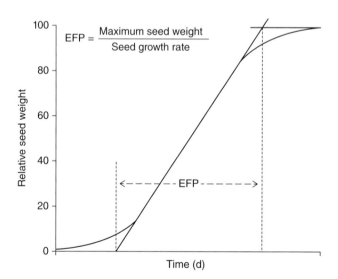

Fig. 3.1. The EFP, calculated by dividing the maximum weight per seed by the seed growth rate during the linear phase of seed growth, represents the time it would take the seed to accumulate its final weight growing at the linear rate. The EFP can be calculated on a community (yield/total seed growth rate) or an individual seed (maximum weight per seed/seed growth rate) basis.

growth curve. As mentioned earlier, some statistical models applied to seed growth are flawed, so estimates of SFD from these models are suspect. Estimates of SFD are dependent upon the method used to produce them; consequently, species or treatment comparisons are valid only when the same technique is used.

Measures of seed characters at the individual seed level require repeated samples of seeds of the same age, i.e. developing from flowers that were pollinated at the same time. This can be accomplished in some crops (e.g. maize or wheat), as discussed in Chapter 2, by sampling a constant position in the inflorescence on plants that began reproductive growth at the same time. In other crops (e.g. soybean and other grain legumes), the location of a flower or fruit on the plant is not necessarily related to when it reached anthesis, making it necessary to identify a group of individual flowers or young fruits that reached a specific growth stage at the same time for later sampling.

Species and Cultivar Variation

Data describing the seed growth characteristics of 14 crop species were collected from the literature to evaluate cultivar and species variation in SGR, SFD and seed size (Table 3.1). These data were collected in environments ranging from uncontrolled conditions in the field to the precisely controlled environment of a phytotron. The data do not represent an exhaustive summary of the literature on seed growth characteristics; the intent was to present a representative sampling of the major grain crops. Data from the more recent literature was added to the summary in Egli (1981) to provide a more representative sampling of all crops. This summary includes 12 of the 22 species in Table 1.1 and the top five species (maize, rice, wheat, soybean and barley) in total world production in 2011 to 2014. A total of 157 genotypes are represented but the distribution of genotypes varies by species, ranging from 32 genotypes for maize to only two each for groundnut and flax (Table 3.1).

There is nearly a 200-fold variation in mean SGR among the species in Table 3.1, with means ranging from 0.2 to 36.9 mg seed^{-1} day^{-1}. Species comparisons are confounded with possible environmental effects; however, many of the species differences are much larger than expected from any possible response to environmental conditions, suggesting that seeds of some species grow faster than those of other species. The two- to fivefold variation in SGR within species suggests that there may also be real differences among cultivars. The SGR of legumes was generally higher than the cereals, with maize, whose SGR was much higher than the other cereals (barley, rice, sorghum and wheat), providing the only exception. Seeds of the three broad bean cultivars and some of the bean cultivars had exceptionally high rates, with one broad bean cultivar reaching 55 mg seed^{-1} day^{-1} (DeKhuijzen and Verkerke, 1986).

The mean EFP of 64% of the species was between 25 and 35 days with only two (13%) less than 20 days; this variation is much less than the 200-fold variation in SGR. Almost all species exhibited some cultivar variation in EFP. The

Table 3.1. Seed growth characteristics of important grain crops.[1]

Species	Number of cultivars	Seed growth rate Mean (mg seed⁻¹ day⁻¹)	Range (mg seed⁻¹ day⁻¹)	EFP[2] Mean (day)	Range (day)	Maximum size Mean (mg seed⁻¹)	Range (mg seed⁻¹)
Cereals:							
Wheat (*Triticum aestivum* L.)	26	1.4	2.1–1.0	29	45–19	41	55–23
Barley (*Horedum vulgare* L.)	13	1.6	2.4–0.6	25	43–18	38	50–22
Rice (*Oryza sativa* L.)	12	1.2	2.0–0.9	24	36–12	28	50–20
Sorghum (*Sorghum bicolor* (L.) Moench.)	9	0.9	1.9–0.4	23	42–20	28	37–19
Maize (*Zea mays* L.) (inbreds)	22	7.4	9.7–3.6	31	39–23	228	322–86
(hybrids)	10	8.8	10.4–7.0	35	41–23	302	410–229
Legumes:							
Soybean (*Glycine max* (L.) Merr.)	21	6.8	14.7–3.6	29	46–13	202	484–84
Bean (*Phaseolus vulgaris* L.)	20	18.9	33.1–10.2	18	24–14	345	540–190
Pea (*Pisum sativum* L.)	5	10.5	14.3–5.6	22	35–12	195	224–150
Field pea (*P. arvense*)	2	9.5	13.0–6.0	25	32–18	211	232–190
Broad bean (*Vicia faba* L.)	3	36.9	55.0–20.0	31	57–16	1104	2017–414
Cowpea (*Vigna unguiculata* L. Walp)	3	8.4	12.2–4.4	8	9–7	73	122–32
Groundnut (*Arachis hypogaea*)	2	12.8	14.0–11.6	44	45–43	563	626–500
Oil seeds:							
Flax (*Linum usitatissimum* L.)	2	0.2	0.3–0.2	31	35–27	8	8–7
Sunflower (*Helianthus annuus* L.)	7	1.6	2.0–1.2	34	48–30	54	75–39

[1]Adapted from Egli (1981) with additional data from the following sources. Wheat: Nichols *et al.* (1985), Schnyder and Baum (1992); barley: Scott *et al.* (1983); rice: Kato (1986), Yang *et al.* (2001); Sorghum: Heinrich *et al.* (1985), Kiniry (1988); maize: Jones and Simmons (1983), Quattar *et al.* (1987), Tollanaar and Bruulsama (1988); soybean: Obendorf *et al.* (1980), Swank *et al.* (1987); bean: Sexton *et al.* (1994); broad bean: Dekhuijzen and Verkerke (1986); groundnut: Sung and Chen (1990); sunflower: Villalobos *et al.* (1994).
[2]Effective filling period.

exceptionally short EFP of bean (18 days) was the mean of 20 genotypes, while the cowpea estimate (8 days) was based on only three genotypes. There are other reports of short EFPs (6–13 days) for cowpea (Wien and Ackah, 1978), so it's possible that short-filling periods are typical for these two species.

Excluding the very small seed of flax, mean seed size exhibited a nearly 40-fold variation among species. The cereals, barley, rice, sorghum and wheat had the smallest seeds (with the exception of flax), while maize seeds were similar in size to many of the legumes. All of the very large seeds (> 500 mg) were produced by legumes, with one genotype of broad bean producing seeds in excess of 2000 mg, but other legumes produced seeds that were relatively small (e.g. cowpea < 100 mg seed^{-1}).

It should be noted that some of these sizes are the mean size of all seeds harvested from a plant or a plant community, but the mean size represents a population of seeds of varying sizes. This variation in size has been documented for most crops (see, for example, rice, Kato, 1986; maize, Tollenaar and Daynard, 1978; wheat, Acreche and Slafer, 2006; sunflower, Unger and Thompson, 1982; groundnut, Sung and Chen, 1990; oat, Doehlert *et al.*, 2008; soybean, Egli *et al.*, 1987c, Egli, 2012). The largest seeds from a soybean community were often more than twice as large as the smallest seeds (Egli *et al.*, 1987c). Substantial differences in mean size were frequently due entirely to changes in the proportions of the various sized seeds with no change in the absolute sizes.

Most of the large variation in seed growth characteristics among species, especially SGR and maximum size, is a genetic characteristic of the species and is not due to environmental conditions. For example, the size and growth rate of a wheat seed will always be less than a maize seed, regardless of environmental conditions. Environmental conditions, however, could be responsible for small differences in growth characteristics, especially in EFP, within and among species, although, as discussed later in this chapter, all seed growth characteristics are under genetic control. Since the data in Table 3.1 represent a summary of many experiments, genetic and environmental effects are completely confounded.

Interestingly, the substantial variation in seed growth characteristics illustrated in Table 3.1 is not related to any of the defining characteristics of the species included in the table. Crops with C4 photosynthesis are generally more productive and have higher crop growth rates than C3 species (Montieth, 1978), but the seeds of C4 species do not necessarily grow faster nor are they larger than those of C3 species (e.g. C4 maize vs C3 legumes). The type of seed seems to be immaterial, with the caryopsis of the cereals producing large and small seeds that grow both fast and slow. Seeds that contain high concentrations of oil and/or protein and require more assimilate to produce 1 g of seed grow as fast or faster and get larger than high-starch (cereal) seeds requiring less assimilate per unit weight (Sinclair and de Wit, 1975). The EFP was relatively stable across all species in Table 3.1, regardless of their characteristics. Overall, our failure to relate seed growth to any common plant characteristics is our first indication that the growth of the seed is, in part, controlled by the seed. The seed is not simply a passive receptacle that is filled by assimilate from the mother plant. This concept will be developed in greater detail in the rest of this chapter and will play a key role in the involvement of the seed in the yield production process.

Variation in SGR within and among species was associated with variation in seed size in the data summarized in Egli (1981) (Fig. 3.2, $r = 0.81**$, $n = 90$). A similar relationship was reported by Lush and Evans (1981). Perhaps it is not surprising that large seeds generally grow rapidly and small seeds grow slowly. The combination of a large seed and a low SGR would require an exceptionally long SFD, longer than could be accommodated in many environments where the time available for plant growth is limited by temperature or the availability of water. Producing a mature seed in a short time (combination of a relatively high SGR and a small size, e.g. cowpea, Table 3.1) may increase the survival potential of the plant in a stressful environment, but this combination would not, as we shall discuss later, be conducive to high yield.

Not all variation in seed size is due to variation in SGR, seed size is also correlated with EFP ($r = 0.50**$, $n = 90$, Egli, 1981). Variation in size at a constant SGR (i.e. size variation is due to variation in EFP) can be found within and between species in Fig. 3.2, but it is less likely than the association between size and rate. The large seed of groundnut is a result of a relatively modest SGR and a long EFP (Table 3.1), while one cowpea genotype represents the other extreme, utilizing a high SGR (12.2 mg seed^{-1} day^{-1}) to produce a relatively small seed (110 mg seed^{-1}) in only nine days. Clearly, SFD contributes to seed size variation, but its contribution is less than SGR. Assuming that large seeds have high SGRs will be correct more often than not.

Genetic (species or cultivar differences) and environmental variation in seed size are determined by SGR and SFD. Understanding the effects of the environment on SGR and SFD and the regulation of these seed growth components will help us understand the relationship between these growth characters, seed size and yield.

Seed Growth Rate (SGR)

The substantial variation in SGR illustrated in Table 3.1 represents 14 crop species grown in many different environments, so it could reflect variation in environmental conditions or it could be an indicator of genetic control. It seems unlikely, however, that the large consistent differences among species would be due to variation in environmental conditions.

Genetic variation

Cultivar differences in SGR that are consistent across environments provide direct evidence of genetic variation. Modification by direct selection also demonstrates genetic control. There is substantial evidence in the literature supporting both approaches. Data from four soybean cultivars growing in two field environments

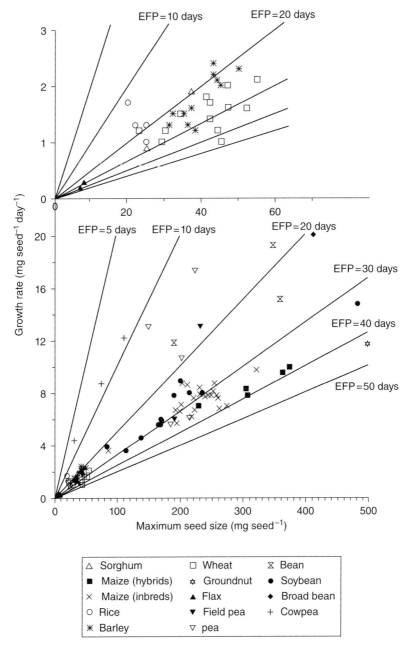

Fig. 3.2. The relationship between SGR, maximum weight per seed and EFP for 13 grain crop species. Each point represents a single cultivar. One broad bean cultivar with a maximum weight seed^{-1} of 2017 mg and a SGR of 35.5 mg seed^{-1} day^{-1} and a bean cultivar with a maximum weight of 480 mg and a rate of 33.1 mg seed^{-1} day^{-1} are not included. Adapted from Egli (1981).

(Table 3.2) illustrate consistent cultivar differences across environments as did the
results of Egli *et al.* (1981). Similar results have been reported for maize (Carter
and Poneleit, 1973), pearl millet (Fussell and Pearson, 1978), wheat (Jenner and
Rathjen, 1978) sorghum (Kiniry, 1988) and rice (Yoshida and Hara, 1977; Fujita
et al., 1984).

Hartung *et al.* (1989) used phenotypic recurrent selection to develop high and
low SGR types in maize. Changes in SGR after three cycles of selection were
4.8 and –8.0% for the high and low selections and the associated heritabilities
were 0.44 for high SGR and 0.20 for low SGR. Heritabilities of SGR in soft red
winter wheat varied from 0.66 to 0.89 (May and Van Sanford, 1992; Mou and
Kronstrad, 1994). Davies (1975) evaluated reciprocal crosses of pea genotypes
that differed in seed size and concluded that the genotype of the seed played a
role in determining SGR.

Evidence from many crops supports the conclusion that SGR is under genetic control and it seems likely that genetic control is a general phenomenon for
all crop species. This evidence for genetic control of SGR supports the previous
suggestion that some of the large differences in SGR in Table 3.1 are not due to
environmental effects, but represent a more fundamental difference among cultivars and species. How these genetic differences are regulated will be discussed
later in this chapter, but the large species differences within and among species
in Table 3.1 and the lack of any relationship between known plant characteristics and SGR suggests that genetic differences in SGR could be determined by a
mechanism intrinsic to the seed.

Environmental and physiological variation

Assimilate supply
The seed cannot grow without a supply of raw materials from the mother plant.
It is not surprising then that SGR responds to assimilate supply as shown in *in vitro*
experiments with soybean (Fig. 3.3 and Egli and Bruening, 2001). Seed growth
rate increased rapidly as the sucrose concentration increased from 0, reaching
90% of the maximum rate at 115 mM when grown with excess supplies of N. Egli

Table 3.2. Seed growth rates of four soybean cultivars growing in the
field for two years. Adapted from Egli *et al.* (1978a).

Cultivar	1974 (mg seed^{-1} day^{-1})	1975 (mg seed^{-1} day^{-1})
Kanrich	6.8[1]	9.1[1]
Williams	5.6	6.2
Cutler 71	5.0	6.1
Essex	3.6	3.7
LSD(0.05)	0.6	0.5

[1]Average of seed growth rates from first and last pods to develop on the plant.

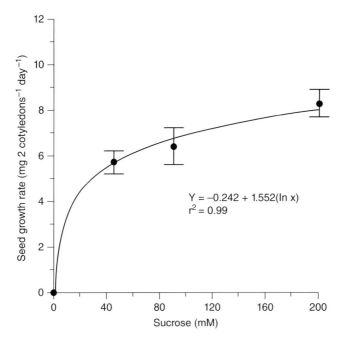

Fig. 3.3. Effect of sucrose concentration on *in vitro* soybean SGR with a complete nutrient media. Cotyledons were cultured for six days and each data point is the mean of ten replicates ± standard error of the mean. Adapted from Egli *et al.* (1989).

and Bruening (2001), using seeds collected early and late in the seed-filling period, found that the growth rate reached 90% of the maximum at sucrose concentrations of 60–120 mM. Similar *in vitro* increases in SGR occurred in wheat with an ear culture system (e.g. Jenner *et al.*, 1991) and in maize where SGR reached a maximum at 60–80 mM sucrose (Cobb *et al.*, 1988), concentrations slightly lower than those reported for soybean. These responses of SGR to sucrose supply are probably characteristic of all grain crop species.

Seeds also require an organic source of N for normal growth, but dry matter accumulation is not always sensitive to the supply of N. For example, removing N from the media in a hydroponics system early in seed filling had no effect on soybean SGR (Fig. 3.4). Hayati *et al.* (1996) found that dry matter accumulation of soybean seeds *in vitro* was relatively insensitive to N levels in the media, with only 17 mM required to maintain normal rates. Seed N concentration was, however, much more sensitive to media N levels, reaching a maximum at 270 mM. They suggested that dry matter accumulation could be maintained with only enough N in the media to maintain metabolic enzymes in the seed. Similar results were reported for pea by Lhuillier-Soundele *et al.* (1999). *In vitro* SGR of maize (Singletary and Below, 1989) and wheat (Barlow *et al.*, 1983) were also relatively insensitive to N supply.

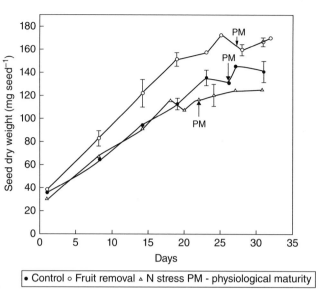

Fig. 3.4. Effect of approximately 75% fruit removal and N stress on soybean seed growth. Nitrogen stress was applied by removing all N from the nutrient media of non-nodulated plants growing in a hydroponic system. Treatments were applied at zero days. Bars represent ± 1 standard deviation. Adapted from Egli *et al.* (1985a).

Jenner *et al.* (1991) concluded that the normal N supply to a wheat seed resulted in apoplastic N concentrations that were on the relatively linear portion of the protein accumulation response curve, suggesting that seed N concentration would be very responsive to seed N supply, as also noted by Hayati *et al.* (1996). It seems then that variation in the supply of N to the seed may have minimal effects on the ability of the seed to accumulate dry matter, but significant effects on the accumulation of N by the seed and hence the seed N or protein concentration. Developing seeds accumulate storage carbohydrates, oil and protein, and it is clear that C and N metabolism are not tightly linked. The ratio of C to N in the raw materials supplied to the seed must be relatively constant to account for the relative stability of seed protein levels among environments. The ability of the seed to accumulate dry matter with only minimal supplies of N suggests that describing legume seeds as having a large N 'demand' (Sinclair and de Wit, 1975) may be incorrect. A better model is one where seeds have no control over the N supply and simply subsist on the N supplied by the plant.

The location of the seed on the plant (i.e. position on the stem or branches or location in the fruit, capitulum, raceme or inflorescence) could affect SGR if there was variation in assimilate supply among locations. Assimilate supplies to soybean fruits located at nodes whose leaves are shaded by other leaves may be less than fruits in higher solar radiation regimes at the top of the canopy. A tip seed in maize is located further from the source of assimilate than a basal seed, and the same can be said for seeds at the top of the cereal inflorescence or outer florets in a spikelet.

The adjustment of seed number to the assimilate supply would, however, tend to maintain a constant supply per seed, minimizing variation in SGR among locations on the plant. The variation of soybean fruits and seeds among main stem nodes of soybean (Egli, 2015a) may be an example of such an adjustment.

When the fruit or seed develops relative to other fruits and seeds on the plant could also influence its assimilate supply, with early developing seeds having a possible advantage. Timing of fruit and seed development is frequently confounded with location, for example, most fruits on the lower nodes of a soybean plant begin development before fruits at upper nodes (there is often 30 to 40 days between appearance of the first and last fruits). The first fruit at an individual node developed up to 18 (indeterminate growth habit) to 34 (determinate growth habit) days before the last fruit (Egli and Bruening, 2006a). Such time differences exist in most crop species; examples include four to five days between tip and basal seeds in maize ears (Tollenaar and Daynard, 1978; Bassetti and Westgate, 1993a), four days among seeds in a wheat head (Evans *et al.*, 1972) and 11 to 12 days in an oat panicle (Rajala and Peltonen-Sarnio, 2004). Such territorial and time advantages for some seeds exist in all grain crop species, and, if these advantages are associated with greater or lesser assimilate supplies, SGR could be affected.

The effect of position or time of development on SGR seems to vary among species. Researchers have reported mixed results for soybean, with Egli *et al.* (1978a) and Wallace (1986) reporting little difference in SGR due to time of flower development or position on the main stem or branches. Gbikpi and Crookston (1981), however, reported that seeds from the first fruits to develop had lower SGR than seeds developing two weeks later, while Yoshida *et al.* (1983) reported that the first fruits had higher SGRs. Relationships are clearer for other species. For example, SGR in wheat clearly depends on the position of the seed in the inflorescence with earlier developing seeds in the basal florets of a spikelet having a higher SGR than those from the more distal tip florets (Rawson and Evans, 1970). Seeds from the basal florets of spikelets at the base of the inflorescence grew more slowly than seeds in the same position in spikelets in the centre of the inflorescence. Seeds at the tip of maize ears frequently grow more slowly than seeds at the base (Frey, 1981). Sorghum seeds from basal flowers had lower SGRs than seed from apical flowers that reached anthesis four to ten days before the basal flowers (Gambin and Borras, 2005). This positional variation, or the lack thereof, is probably due to variation in the supply of assimilate during the linear phase of growth, but effects of the environment or the supply of assimilate on the characteristics and growth potential of the individual seed cannot be discounted. The variation in SGR resulting from positional or timing effects could be responsible for some of the intra-plant variation in seed size discussed earlier in this chapter.

Water stress

The effect of water deficits on all aspects of plant growth and yield are well documented. Water stress reduces stomatal conductance and photosynthesis in the leaf and could, at least in theory, affect the metabolic capacity of the seed. There is, however, little evidence that moderate water stress has any direct effect on SGR.

Short-term stress that was shown to reduce leaf water potential had no effect on SGR in soybean (Westgate *et al.*, 1989; Westgate and Thomson Grant, 1989a), maize (Westgate and Thomson Grant, 1989b), pea (Ney *et al.*, 1994), barley, and wheat (Brooks *et al.*, 1982). Long-term water stress in the field or in the greenhouse reduced yield but had no effect on SGR in soybean (Meckel *et al.*, 1984; Egli and Bruening, 2004) and maize (Quattar *et al.*, 1987). Severe stress, however, reduced SGR in barley (Aspinall, 1965) and, when applied very early in seed development, affected metabolic activity of maize seeds (Artlip *et al.*, 1995; Zinselmeier *et al.*, 1995).

A constant SGR during stress suggests that the supply of assimilate to the seed is not reduced or not reduced enough to affect SGR (see Fig. 3.3) and the ability of the seed to metabolize incoming assimilate is not compromised. As discussed previously, reductions in photosynthesis do not necessarily affect the supply of assimilate to the developing seed (timing of the reduction is critical). Mobilization of reserve assimilates (e.g. Quattar *et al.*, 1987) may help maintain a relatively constant supply per seed. If the stress significantly reduces the supply of assimilate to the seed or if the metabolic capacity of the seed is impaired, SGR could be reduced by water stress.

There is little evidence that seed water potential changes when the plant is stressed, so the metabolic capacity of the seed may not be reduced by water stress. Constant seed water potentials during stress episodes causing decreases in leaf water potential have been documented in several crops (Shackel and Turner, 2000; see Bradford (1994) for a review). Severe stress, however, affected seed water potential and SGR in maize (Artlip *et al.*, 1995; Zinselmeier *et al.*, 1995). The stability of seed water potential is usually explained by the lack of a vascular connection between the developing embryo and endosperm, and the mother plant (Saini and Westgate, 2000). The resulting discontinuity provides the resistance that allows the differences in water potential to exist. This viewpoint was challenged by Bradford (1994), who suggested that the apparent differences between seed and other plant tissue may be artifacts of the techniques used to measure seed water potential. Regardless of this controversy, developing seeds seem to be remarkably resistant to water stress, maintaining SGR under all but the most severe stress. Thus, SGR generally does not play a significant role in the response of the plant to water stress during seed filling. Adjustments in seed size under stress are more likely a result of shortening the SFD than reducing SGR.

Temperature

The metabolic processes that produce plant growth are affected by temperature, so it is not surprising that SGR responds to variation in temperature. There was a linear reduction in SGR of wheat, soybean, rice and maize as temperature decreased below approximately 22°C (Fig. 3.5). The relative decrease in SGR seems to be similar for the four species, even though the experiments were conducted in different environments with different combinations and durations of day/night temperatures. The SGR of sorghum at ~ 20°C is much lower than the other species, but there is only a single observation below the optimum temperature. High temperatures also reduce SGR, although again determination of critical

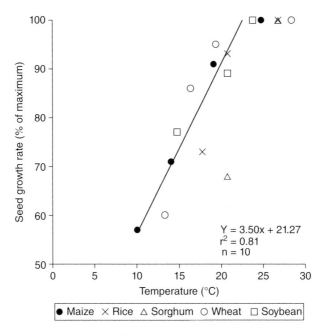

Fig. 3.5. Temperature effects on SGR of several grain crops. Seed growth rates are expressed as a percentage of the maximum rate. The regression analysis did not include the observations at 100% or any of the sorghum data. Data were adapted from Tollenaar and Bruulsema (1988), maize; Chowdhury and Wardlaw (1978), rice and sorghum; Sofield *et al.* (1977), wheat; and Egli and Wardlaw (1980), soybean.

temperatures and comparisons among species are limited by the availability of data. Many experiments have shown reduced SGRs for temperatures above 30°C (e.g. Tashiro and Wardlaw (1989), wheat and rice; Egli and Wardlaw (1980) and Gibson and Mullins (1996), soybean; Jones *et al.* (1981), maize). Tashiro and Wardlaw (1989) found similar optimum temperatures (30/25°C) for wheat and rice, but the SGR of wheat declined much more rapidly than rice at temperatures above the optimum.

Temperature could affect SGR directly by affecting seed metabolism or by affecting the supply of assimilate to the seed. Temperature responses *in vitro* when assimilate supplies were not limiting were usually similar to *in vivo* responses (soybean, Egli and Wardlaw, 1980; maize, Jones *et al.*, 1981; wheat, Donovan *et al.*, 1983). High temperature reduced the incorporation of ^{14}C into starch in the endosperm of wheat (Bhullar and Jenner, 1986). Jones *et al.* (1984) demonstrated that high (35°C) and low (15°C) temperatures during the lag phase of seed development influenced the *in vitro* SGR of maize seeds. Egli and Wardlaw (1980) reported similar results for soybean. These results suggest that temperature effects on SGR are primarily on the metabolic capacity of the seed to accumulate dry matter with lesser effects from variation in the assimilate supply from the mother plant.

Miscellaneous factors

Seeds are rich sources of plant hormones and occasional reports have suggested that these hormones may influence SGR. For example, there were reports that ABA influenced C and N movement into soybean seeds, but they could not be confirmed (see Schussler and Brenner (1989) for a review). Hormone levels often vary substantially during seed development (Bewley *et al.*, 2013, pp. 36–43), while SGR is constant, making it difficult to postulate a major regulatory role for hormones. Photoperiod has also been reported to influence SGR in soybean (Morandi *et al.*, 1988) but this response may be an indirect result of changes in assimilate partitioning, rather than a direct effect on the ability of the seed to accumulate dry matter.

Regulation of seed growth rate

Our previous discussion suggested that SGR could be regulated by the seed or by the mother plant. Seed growth rate could be determined by the capacity of the seed to accumulate dry matter or, alternatively, the seed could simply be a passive receptacle for assimilate from the mother plant, in which case, SGR would be completely regulated by the ability of the plant to supply assimilate.

Assimilate supply

The supply of raw materials for seed growth is ultimately related to photosynthesis, raising the possibility that environmental effects on photosynthesis and the assimilate supply could affect SGR. Of course, if SGR is not limited by the supply of raw materials but by processes in the seed, there should be no relationship between the supply of assimilate and SGR. The environment (e.g. temperature) could also directly affect seed metabolism and the ability of the seed to synthesize storage materials. These categories may not be totally exclusive, but they provide a useful framework to consider regulation of seed growth, one that has important implications for understanding the relationship between source and sink and the yield production process.

Soybean SGR was directly related to sucrose concentrations up to approximately 100 mM in an *in vitro* culture system, but there was little change as the concentration increased to 200 mM (Fig. 3.3). If sucrose concentration in the apoplast is above100 mM, changes in assimilate supply will not affect SGR. Consequently, the sensitivity of SGR to the assimilate supply *in vivo* will be determined by sucrose concentration in the seed. Gifford and Thorne (1985) estimated that *in vivo* sucrose concentration in the apoplast of developing soybean seeds was between 100 and 200 mM. Westgate *et al.* (1989) also reported approximately 100 mM, but Hsu *et al.* (1984) reported lower levels (approximately 35 mM). Fisher and Gifford (1986) reported sucrose concentrations of 60–80 mM in the endosperm cavity of wheat, while Jenner *et al.* (1991) suggested that the sucrose concentration was usually on the saturation part of the response curve. Accurate measurements of apoplastic sucrose concentrations are notoriously difficult, consequently, little is known of species, cultivar or environmental variation. The sensitivity of *in vivo* SGR to environmental conditions and treatments that potentially alter the assimilate supply to the seed may provide some indication of seed sucrose levels. If normal apoplastic sucrose concentrations are greater than ~ 100 mM, SGR should be relatively insensitive to the variation in assimilate supply, but SGR could be more variable if they are routinely less than 100 mM.

The relationship between the assimilate supply and seed growth *in vivo* is usually investigated by manipulating the assimilate supply to the seed by reducing the source (defoliation or shade treatments are common) or by reducing the sink (fruit removal or restricting pollination). In general, the effects of these treatments on SGR are quite variable. Reducing seed number increased SGR of soybean (Fig. 3.4, Egli *et al.*, 1989), wheat (Table 3.3), maize (Borras *et al.*, 2003), and sorghum (Gambin and Borras, 2007b).

Table 3.3. (a and b). Source–sink alterations and seed growth rates in soybean and wheat.

(a) Soybean

	Shade from R1 to PM[1]			Shade from R6 to PM[2]	
Cultivar	Control (mg seed^{-1} day^{-1})	Shade (mg seed^{-1} day^{-1})	Cultivar	Control (mg seed^{-1} day^{-1})	Shade (mg seed^{-1} day^{-1})
McCall	5.5	5.9	Kasota	5.8	4.3
Hardin	4.9	4.4	Hardin	4.9	3.9
Harper	6.2	6.2	Hutcheson	4.3	3.6
Essex	4.5	3.9	Essex	4.0	3.4

(b) Wheat[3]

	Early[4]			Late[4]		
Cultivar	Control (mg seed^{-1} day^{-1})	Seed removal[5] (mg seed^{-1} day^{-1})	Defoliation[6] (mg seed^{-1} day^{-1})	Control (mg seed^{-1} day^{-1})	Seed removal (mg seed^{-1} day^{-1})	Defoliation (mg seed^{-1} day^{-1})
Era	1.5[7]	1.8	1.1	1.2	1.8	1.3
Olaf	1.7	2.0	1.2	1.7	2.0	1.6

[1]Adapted from Egli (1993), 63% shade applied from initial bloom (growth stage R1) to physiological maturity, 1989–1990.
[2]Adapted from Egli (1999), 63% shade applied from early seed filling (growth stage R6) to physiological maturity, 1993–1995.
[3]Adapted from Simmons *et al.* (1982).
[4]Treatments were applied at anthesis (early) or 14 d after anthesis (late).
[5]Removal of approximately 50% of the seeds.
[6]Removal of top four leaves on each culm.
[7]Average of seed from three (Olaf) or four (Era) positions in the spike and two years.

But in other experiments, sink reduction failed to stimulate SGR of soybean (Egli and Bruening, 2001), wheat (Slafer and Savin, 1994), maize (Frey, 1981) and sorghum (Kiniry, 1988).

Reducing assimilate supply during seed filling, however, usually reduces SGR (soybean, Egli *et al.*, 1985a, Egli, 1999, Egli *et al.*, 1989, Egli and Bruening, 2001; wheat, Grabau *et al.*, 1990; sunflower and maize, Andrade and Ferreiro, 1996). There was, however, little variation in soybean SGR among years in the field (Table 3.4), perhaps suggesting that *in vivo* SGR is relatively insensitive to variation in assimilate supply.

It is not clear if this inconsistent response to gross manipulations of source–sink ratios is a result of variation in the effect of the treatment on the supply of assimilate to the seed or variation in the ability of the seed to respond to changes in the assimilate supply. The ability to respond would depend upon the initial level of assimilate in the seed (Fig. 3.3); SGR would respond at low initial levels while higher levels would produce no effect.

Source–sink modification treatments that enhance the source relative to the sink (e. g. seed or fruit removal, pollination restriction) are assumed to increase the supply of assimilate to the individual seed, while treatments that limit the source (e.g. shading or defoliation) are assumed to reduce the supply of assimilate to the individual seed. Jenner (1980) discussed the difficulties inherent in these assumptions and in interpreting the results of source–sink modification experiments; those difficulties stem from having no information on the effect of the treatment on the supply of assimilate to the seed. For example, Jenner (1980) found that shading reduced the supply of sucrose to the wheat endosperm, but seed removal did not increase it.

Source–sink alteration treatments may have mixed effects on SGR, depending on the plant growth stage when they are applied. If the plant responds to a reduction in assimilate (shade or defoliation treatments) by reducing seeds per plant, there may be no change in assimilate supply per seed and therefore no effect on SGR. The assimilate supply per seed is more likely to be affected if the same treatment is applied at a later growth stage, when there is no effect on seed number. An example of the importance of timing is shown in Table 3.3, where shading soybean plants during flowering, pod set and seed filling (growth stage R1

Table 3.4. Seed growth rate of two soybean cultivars in five irrigated field environments.

	Seed growth rate				
Cultivars	1989[1] (mg seed^{-1} day^{-1})	1990[1] (mg seed^{-1} day^{-1})	1993[2] (mg seed^{-1} day^{-1})	1994[2] (mg seed^{-1} day^{-1})	1995[2] (mg seed^{-1} day^{-1})
Hardin	5.4	4.6	5.0	5.5	4.3
Essex	4.8	4.1	3.6	4.3	4.2

[1]From Egli (1993).
[2]From Egli (1999).

to physiological maturity) reduced seed number and had minimal effects on SGR, while shade only during seed filling had limited effects on seed number but consistently reduced SGR by 15–26%.

The tendency of all grain crops to adjust seed number per plant or per unit area to the productivity of the environment (see Chapter 4) minimizes variation in the supply of assimilate to the individual seed and tends to maintain stable seed assimilate levels. Consistent stress that reduces photosynthesis and yield by 50% or more may, for example, have no effect on SGR, because a 50% reduction in seed number maintained the apoplastic sucrose concentration in each seed at a level similar to a non-stressed plant. The substantial fluctuations in seed number in response to environmental conditions (see Chapter 4) suggest that this mechanism plays an important role in stabilizing SGR (Table 3.4) and making it possible for the plant to produce a normal-sized seed in a wide range of environments. Intermittent stress during reproductive growth could, however, upset the balance between the productivity of the community and seed number, causing changes in the supply of assimilate per seed and potentially changes in SGR. This variation will be discussed at length in Chapter 5.

The limited data on apoplastic seed sucrose levels makes it difficult to draw clear conclusions regarding the importance of regulation of SGR by assimilate supply in the field. Reducing the assimilate supply during seed filling seems to reduce SGR more consistently than increasing the supply by artificially reducing seed number. These responses are consistent with the sucrose concentration in the apoplast usually being near the level (~100 mM for soybean in Fig. 3.4) where the SGR saturates, consequently a reduction in assimilate supply would reduce SGR while an increase would have little effect. The exceptions to this scenario in the literature could indicate that concentration varies from this critical level in some environments.

It is not known whether assimilate supply differences are responsible for variation in SGR among locations on the plant or times of development. Direct *in vivo* measurements of assimilate supply to an individual seed are generally not available to answer this question, but it would not be surprising if the location of the seed on the plant or the timing of seed development affected the access of the developing seed to assimilate. Early developing seeds are often located closer to the source and thus could have preferential access to the assimilate supply, leaving less for the late developing seeds. As Evans (1993, p. 236) pointed out: 'as in human affairs, it pays to be large ("early developing"), close to the source and well connected'.

The ability of the seed to respond to variation in the assimilate supply plays an important role in the determination of seed size and the debate over whether source or sink limits yield. We will return to this discussion in Chapter 5.

We have discussed regulation of SGR primarily from the viewpoint of assimilate supplies. Assimilate usually refers only to C, but seeds also require N for growth. Ignoring the N supply when evaluating the regulation of SGR probably does not create serious problems, given the relative insensitivity of SGR to N supply (Fig. 3.4) (Barlow *et al.*, 1983; Singletary and Below, 1989; Hayati *et al.*,

1996). Seed growth rate could be reduced, however, if N stress reduced photo-synthesis and the assimilate supply. Seed N concentration is much more sensitive to changes in N supply than SGR (Hayati *et al.*, 1996; Allen and Young, 2013), so changes in N supply with a constant C supply could affect seed N concentration without any effect on SGR. If C and N, however, maintained a constant ratio as the assimilate supply changed, the seed N concentration may show no change, although the SGR could change. Given the relatively close linkage between N ac-quisition and photosynthesis in many crops, one could speculate that C:N ratios in the assimilate supply may remain relatively constant in normal field environments, although it's also possible that stress could perturb the ratio, leading to changes in seed composition.

Seed characteristics

Seeds depend upon the mother plant for raw materials, so it is not surprising that the supply of sucrose and N affects SGR. The assimilate supply from canopy photosynthesis is determined by species (C_3 vs C_4) and all the environmental con-ditions that influence crop productivity, but the supply to an individual seed is modulated by the relationship between the assimilate supply and the number of seeds, which, as just discussed, may provide a relatively constant supply of assimi-late per seed. Consequently, it is unlikely that variation in assimilate supply per seed is responsible for the large differences in SGR among species (Table 3.1). The characteristics of the seed have a role to play in determining SGR, a role that is independent of the assimilate supply.

Genetic differences in SGR are primarily regulated by the seed through the number of cells in the cotyledons or endosperm. Evidence supporting this mech-anism falls into two categories. First, there are reports for several species of signifi-cant correlations between cell numbers and SGR across genotypes with substantial differences in SGR; such relationships have been reported for soybean (Egli *et al.*, 1981, 1989; Guldan and Brun, 1985; Munier-Jolian and Ney, 1998), maize (Reddy and Daynard, 1983; Jones *et al.*, 1996), wheat (Jenner and Rathjen, 1978), pea and barley (Cochrane and Duffus, 1983). Reddy and Daynard (1983) and Jones *et al.* (1996) also reported a close association between the number of starch grains in maize endosperms and genetic differences in SGR. Positive correlations between the number of cells in the cotyledons and seed size across genotypes of pea and in the genus *Vicia* were reported by Davies (1975, 1977); these differences in seed size were probably associated with differences in SGR. Second, genetic differences in SGR in soybean associated with differences in cotyledon cell numbers were main-tained in *in vitro* culture systems containing excess levels of C and N (Fig. 3.6; Egli *et al.*, 1981, 1989) clearly demonstrating that the differences in SGR were regu-lated by the characteristics of the seed, not by the supply of assimilate to the seed.

Cell division in the cotyledons or endosperm is complete (or nearly so) before the seed enters the linear growth phase, so the 'machinery' for growth (the number of cells per seed and probably the quantity of enzymes per cell) remains constant during the linear phase of growth. The accumulation of dry matter by the seed is a result of synthesis of storage materials, which does not contribute to the growth

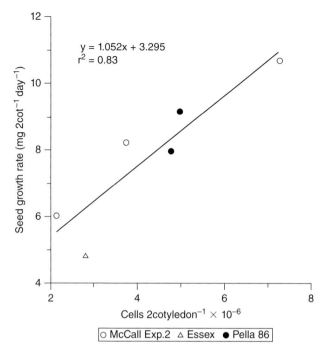

$$y = 1.052x + 3.295$$
$$r^2 = 0.83$$

○ McCall Exp.2 △ Essex ● Pella 86

Fig. 3.6. The relationship between the number of cells per seed and the *in vitro* SGR with excess levels of C and N in the culture media. McCall seeds were collected from plants subject to fruit removal and shade treatments. A control sample was also cultured. From Egli *et al.* (1989).

potential of the seed. The fact that the amount of 'machinery' is constant during the linear phase of seed growth would be expected if it is playing a regulatory role in seed growth.

Regulation by the seed helps explain the large species differences in SGR. The relatively low SGR of a rice or wheat seed does not represent a weakness in the plant's ability to supply assimilate to the seed, nor do the high rates of some legumes (Table 3.1) give any indication of superior photosynthetic capabilities. Instead, these species differences in SGR are regulated by the characteristics of the seed, not by the supply of assimilate to the seed. Soybean seeds have many more cells in the cotyledons ($2–10 \times 10^6$ per seed, Egli *et al.*, 1981, 1989; Guldan and Brun, 1985) than wheat or barley ($0.05–0.15 \times 10^6$ per seed, Wardlaw, 1970; Brocklehurst, 1977; Singh and Jenner, 1982; Djarot and Peterson, 1991) and have higher SGRs ($3.6–10.4$ mg seed^{-1} day^{-1}) than wheat ($1.0–2.1$ mg seed^{-1} day^{-1}) (Table 3.1). Maize represents a conspicuous exception with reported cell numbers similar to wheat and barley ($0.1–0.6 \times 10^6$ per seed, Quattar *et al.*, 1987; Jones *et al.*, 1996) while the SGR is similar to soybean ($3.6–10.4$ mg seed^{-1} d^{-1}). Sunflower is another example of a seed with a large number of cells ($2–2.5 \times 10^6$ cells seed^{-1}) and a modest SGR ($1.2–2.0$ mg seed^{-1} d^{-1}) (Lindstrom *et al.*, 2006).

The metabolic capacity per cell must vary among species, but the basis for this variation is not known. Regardless of the mechanism, it is clear that genotypic differences in SGR within a species and differences among species are regulated by the seed, not by the supply of assimilate from the mother plant. Regulation by the seed provides a mechanism to regulate the number of seeds per plant and we will discuss this in detail in Chapter 4.

Temperature and the assimilate supply during the cell division phase can influence the number of cells in the cotyledons or endosperm. Pod removal or shade treatments applied to alter the supply of assimilate to soybean seeds during the cell-division phase changed the number of cells in the cotyledons and the *in vitro* SGR with non-limiting supplies of C and N (Fig. 3.6, Egli *et al.*, 1989). Low light levels also reduced endosperm cell number in wheat (Wardlaw, 1970) and sunflower seeds (Lindstrom *et al.*, 2006). Temperature affected the rate of cell division in wheat (Wardlaw, 1970) but did not affect the final cell number; however, high temperatures reduced cell number in maize endosperms (Jones *et al.*, 1985). Moisture stress during the early stages of seed development reduced cell number in wheat and maize (Brocklehurst, 1977; Quattar *et al.*, 1987; Artlip *et al.*, 1995) and *in vitro* C and N levels affected cell number in detached cultured wheat ears (Singh and Jenner, 1984). Swank *et al.* (1987) reported variation in cotyledon cell number of a single soybean genotype between years, providing some evidence that cell numbers can respond to the environment in the field. Weber *et al.* (1998) provided a link between assimilate supply and cell division by suggesting that high hexose to sucrose ratios might favour cell division and increase cell numbers. The magnitude and frequency of these environmental effects in the field is not well documented, but they occur and their effect on the yield production process should be similar to the effect of genetic variation in cell number.

The environment could also modulate SGR by directly affecting the capacity of the seed to synthesize storage reserves. Temperature control of metabolic rates in the seed represents, in a sense, a temporary change in seed characteristics that affects SGR. There is little evidence that water stress directly affects the seed. Other environmental factors, such as solar radiation, probably affect SGR by modifying the supply of assimilate to the seed.

Summary

Seed growth rate can be regulated by the supply of assimilate from the mother plant and/or by the capacity of the seed to synthesize storage compounds. *In vitro* culture systems made a significant contribution to our understanding of these two components by providing a means of directly manipulating the supply of C or N to the developing seed. These systems made it possible to separate the role of seed characteristics from control by the supply of assimilate from the mother plant.

Control by the seed includes genetic differences in SGR and direct effects of the environment on metabolic processes in the seed. Seeds cannot grow without a constant supply of assimilate from the mother plant, which provides another mechanism to regulate SGR. The adjustment of the number of seeds to the availability of assimilate probably maintains the assimilate supply per seed at a relatively constant level, minimizing effects of variation in assimilate supply. Control of SGR by the seed provides a basis for understanding how plants determine how many seeds to produce, the relationship between SGR, seed size and yield, and source and sink limitations of yield. We will investigate these relationships in the following chapters.

Seed-Fill Duration (SFD)

The final size of the seed that is harvested for yield is a function of both SGR and SFD. Although much of the variation in seed size is associated with SGR, variation of SFD may also contribute to variation in seed size. Seed-fill duration cannot be ignored as we investigate the involvement of the seed in the determination of yield.

Genetic variation

Genotypic differences in SFD that are consistent across years and environments have been found in many crops, suggesting that SFD is under genetic control. Such evidence has been reported for soybean (Table 3.5, Hanway and Weber, 1971; Gay et al., 1980); maize (Daynard et al., 1971; Poneleit and Egli, 1979), wheat (Chowdhury and Wardlaw, 1978; Gebeyehou et al., 1982; Van Sandford, 1985), rice (Chowdhury and Wardlaw, 1978; Kato, 1999), barley (Metzger et al., 1984; Garcia del Moral et al., 1991; Leon and Geisler, 1994; Dofing, 1997), oat (Wych et al., 1982; Peltonen-Sainio, 1993), sorghum (Sorrells and Meyers, 1982), sunflower (Villalobos et al., 1994) and common bean (Sexton et al., 1994). Maize hybrids had longer SFD than inbreds (Johnson and Tanner, 1972; Poneleit and Egli, 1979).

Seed-fill duration was modified by direct selection in winter (Mou and Kronstrad, 1994) and spring wheat (Talbert et al., 2001), barley (Rasmusson et al.,

Table 3.5. Genotypic differences in soybean seed-fill duration in three environments.

Genotype	Seed-filling period[1]		
	1976[2] (days)	1978[3] (days)	1979[4] (days)
Williams	48	41	39
Lincoln	43	27	30

[1]Growth stage R4 to R7 in 1976, growth stage R5 to R7 in 1978 and 1979.
[2]Gay et al. (1980).
[3]Zeiher et al. (1982).
[4]Boon-Long et al. (1983).

1979; Metzger *et al.*, 1984), soybean (Metz *et al.*, 1985; Smith and Nelson, 1986b; Pfeiffer and Egli, 1988) and maize (Cross, 1975; Fakorede and Mock, 1978). Hartung *et al.* (1989) increased SFD by 4.8 days (15%) with three cycles of recurrent selection in maize, while Smith and Nelson (1986b) developed F5 soybean lines with SFDs averaging three days (7%) longer than lines selected for short SFD.

Estimates of heritabilities of SFD in soybean ranged from –0.20 to 1.02 (Metz *et al.*, 1984, 1985; Salado-Navarro *et al.*, 1985; Smith and Nelson, 1987; Pfeiffer and Egli, 1988). Mou and Kronstrad (1994) reported heritabilities of 0.84 for wheat, but estimates for maize, when selecting for long and short SFDs, were much lower (0.19 and 0.14) (Hartung *et al.*, 1989). Heritabilities for barley were as high as 0.94 (Rasmusson *et al.*, 1979).

Plant breeders also inadvertently lengthened the seed-filling period when selecting for higher yield. Modern maize hybrids have longer seed-filling periods than older hybrids (McGarrahan and Dale, 1984; Frederick *et al.*, 1989; Bolanos, 1995; Duvick, 2005). This advantage for modern hybrids was apparent in a wet year with high yield and a dry year with lower yields (Fig. 3.7), so the extended period seemed to be independent of stress. Breeding for higher yield also lengthened the seed-filling period in groundnut (Duncan *et al.*, 1978), oat (Peltonen-Sainio, 1993) and soybean (McBlain and Hume, 1981; Wells *et al.*, 1982; Shiraiwa and Hashikawa, 1995). Domestication increased the seed-filling period in wheat (Evans and Dunstone, 1970) and maize (Gardner *et al.*, 1990). The EFP of *Glycine soja*, a wild relative of cultivated soybean, was 22 days compared with 30 days for commercial *G. max* cultivars (average of 18 cultivars with a range of 24–34 days, Egli, unpublished data, 1998). A longer SFD could also be associated with the stay-green characteristic; a canopy characteristic that has been associated with genetic yield improvement in maize (Duvick, 2005) and sorghum (Monk *et al.*, 2014).

The evidence that SFD is under genetic control is compelling and comes from work with many important grain crops. The substantial variation within species in Table 3.1 suggests that genetic variation in SFD is a common feature of all grain crops.

Environmental and physiological variation

Temperature

The SFD of most crops is affected by temperature. Seed-fill duration, estimated by EFP, generally increased as temperature decreased in a summary of data from the literature representing four important crops (Fig. 3.8).

Decreasing temperature from 30° to 20°C more than doubled SFD (14 to 29 days) for these species. In contrast, SFD of soybean (Hesketh *et al.*, 1973; Egli and Wardlaw, 1980) and rice (Chowdhury and Wardlaw, 1978) were relatively insensitive to temperatures between 20 and 30°C. Temperatures above 30°C would result in even shorter seed-filling periods. The CROPGRO simulation model produced improved predictions of soybean yield in cool climates when

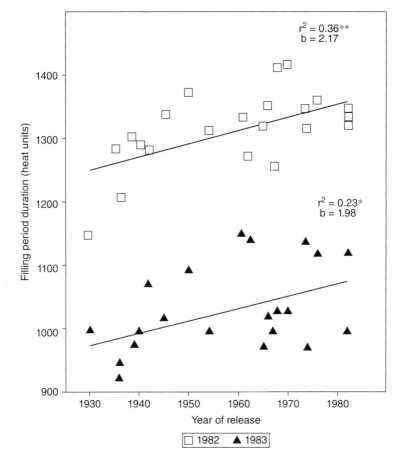

Fig. 3.7. Seed-filling duration of maize hybrids released between 1936 and 1982 in the USA. Includes one open pollinated genotype from 1930. Slopes of the linear regression equations were not significantly different. From Cavalieri and Smith (1985).

the relative sensitivity to temperature during seed filling was less than during early reproductive growth (Boote *et al.*, 1998). The available data, especially species comparisons in the same experiment (e.g. Chowdhury and Wardlaw, 1978), suggest that there may be species differences in the sensitivity of SFD to temperature. Since SFD is an important determinant of yield (discussed in Chapter 5), species differences in sensitivity could be important in adapting to a warming world.

Water stress

Water stress during seed filling shortens the seed-filling period. Severe stress during seed filling caused physiological maturity to occur earlier (18–29%), shortening the

SFD and reducing yield (26–44%) and seed size (7–32%) in soybean (Table 3.6). This response is typical of other species such as barley (Aspinall, 1965; Brooks *et al.*, 1982), wheat (Brooks *et al.*, 1982), rice (Yang *et al.*, 2001), maize (Jurgens *et al.*, 1978; Quattar *et al.*, 1987), pearl millet (Bieler *et al.*, 1993) and chickpea

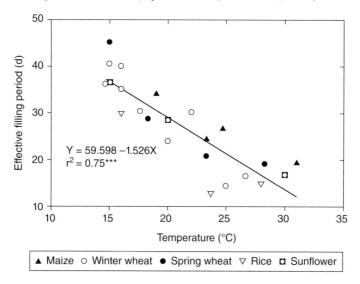

Fig. 3.8. The relationship between temperature and seed-fill duration for several crop species. Seed-fill duration was estimated by the effective filling period and data from each source was averaged across genotypes, years or experiments where appropriate. The regression was significant at *p* < 0.001. Maize – Tollenaar and Bruulsema (1988), Wilhelm *et al.* (1999); wheat – Vos (1981), Tashiro and Wardlaw (1989), Hunt *et al.* (1991), Wardlaw and Moncur (1995); rice – Fujita *et al.* (1984), Tashiro and Wardlaw (1989); sunflower – Chimenti *et al.* (2001). From Egli (2004).

Table 3.6. Water stress during seed filling and duration of seed fill in soybean in the greenhouse. From de Souza *et al.* (1997).

					Physiological maturity[2]	
	Yield		Seed size		Exp. 1	Exp. 2
Moisture level[1]	Exp. 1[3] (g plant⁻¹)	Exp. 2[3] (g plant⁻¹)	Exp. 1 (mg seed⁻¹)	Exp. 2 (mg seed⁻¹)	(days after R6[1])	(days after R6[1])
Well watered	17.0	39.5	128	240	22	24
Moderate stress	17.2	31.6	143	207	19	20
Severe stress	12.5	22.1	119	163	18	17
LSD(0.05)	2.6	2.6	11	18	1	1

[1]Treatments were applied at growth stage R6.
[2]Growth stage R7, at least one mature pod on the main stem.
[3]Cultivar McCall in Exp. 1 and Elgin 87 in Exp. 2.

(Davies *et al.*, 1999). Moisture stress during seed filling also accelerates leaf senescence (maize, Jurgens *et al.*, 1978, Aparicio-Tejo and Boyer, 1983; soybean, de Souza *et al.*, 1997; Brevedan and Egli, 2003, Egli and Bruening, 2004; chickpea, Davies *et al.*, 1999; sunflower, Whitfield *et al.*, 1989) which should lead to a shorter seed-filling period. Brevedan and Egli (2003) reported that stress-triggered accelerations of senescence were not reversed when plants were returned to well watered conditions after only three days of stress (Fig. 3.9). Photosynthesis was quickly restored to the control level, but the acceleration of senescence continued, reducing yield by 17% (vs 39% in the continuous stress). Since seeds cannot grow without raw materials from the mother plant, the premature decline in the assimilate supply as a result of the acceleration of leaf senescence is probably the primary cause of the shorter SFD. This point will be discussed in greater detail when we consider the regulation of SFD. Water stress during seed filling represents, in some respects, a hidden stress in that the senescence process seems entirely normal; only comparison to a well watered control reveals that it is occurring sooner. If only a few days of water stress are required to accelerate senescence, stress may limit yield in what seem to be relatively well watered environments. This scenario suggests that a complete absence of water stress during seed filling may be required for maximum yield.

Assimilate supply

The effects of variation in the assimilate supply on SFD are more complex than they are on SGR. Both processes depend upon the mother plant for a supply of raw materials, but, in the case of SFD, how long the assimilate supply is maintained is key whereas SGR is related to the rate of supply (assimilate per day). The capacity of the seed to respond to an extended supply of assimilate is also important. The effects of assimilate supply on SFD revolve around two questions: how long is the canopy photosynthetically active and how long can the seed continue growth?

Shortening the period when assimilate is available obviously shortens the seed-filling period, because seeds cannot grow without a source of raw materials. Complete defoliation shortened the seed-filling period in maize (Jones and Simmons, 1983; Hunter *et al.*, 1991), sorghum (Rajewski and Francis, 1991) and soybean (Vieira *et al.*, 1992).

A variety of stresses during seed filling may reduce the time that assimilate is available to the developing seed, shortening the seed-filling period. Examples include leaf disease in wheat (Pepler *et al.*, 2005), nutrient stress in maize (Peaslee *et al.*, 1971), N stress in soybean (Fig. 3.3) and water stress in several crops as discussed previously. Partial defoliation or shade treatments designed to produce only modest reductions in assimilate did not consistently affect SFD. Shade treatments that reduced irradiance by 45 to 63% had no effect or lengthened SFD in soybean (Egli *et al.*, 1985a; Andrade and Ferreiro, 1996; Egli, 1999). The seed-filling period of sunflower was shortened by 45% shade in one of two years, but this treatment had no effect on maize (Andrade and Ferreiro, 1996). Partial defoliation shortened SFD in grain sorghum (Rajewski and Francis, 1991) and in one experiment with

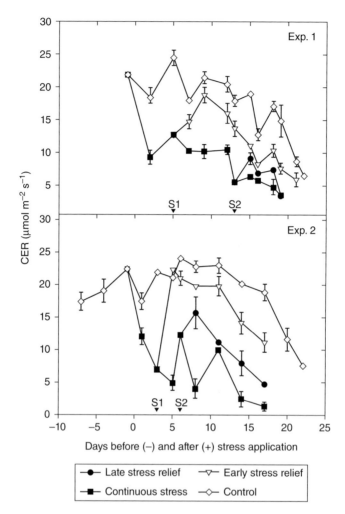

Fig. 3.9. Effect of water stress on carbon exchange rate (CER) during seed filling of soybean in two greenhouse experiments. Stress was applied early in seed filling at the beginning of growth stage R6. The time of application of the stress-relief treatments (plants were returned to well watered controls) is shown on the x-axis by S1 (early) and S2 (late). Bars represent ± one standard error of the mean. Some error bars were omitted to avoid excessive clutter. From Brevedan and Egli (2003).

soybean (Munier-Jolain *et al.*, 1998) but not in others (Egli and Leggett, 1976) or maize (Frey, 1981).

Pod or seed removal to increase the supply of assimilate to the remaining seed lengthened the seed-filling period of soybean (Konno, 1979; Egli *et al.*, 1985a; Munier-Jolain *et al.*, 1996; Egli and Bruening, 2001), but not maize (Jones and Simmons, 1983; Kiniry, 1988) or wheat (Slafer and Savin, 1994). Reducing seed number slowed leaf senescence of maize in three of four comparisons (Borras *et al.*, 2003). Reducing plant density at the beginning of the seed-filling period to increase photosynthesis per plant had no effect on SFD of maize or soybean, but increased it in sunflower (Frey, 1981; Andrade and Ferreiro, 1996). Increasing plant density accelerated leaf senescence in maize (Poneleit and Egli, 1979; Borras *et al.*, 2003) and shortened the seed-filling period. Exposing plants to atmospheres enriched with CO_2 did not affect SFD in wheat (Wheeler *et al.*, 1996) or lupin (lupines albus L., Munier-Jolian *et al.*, 1998). Seed-fill duration did not respond to increased irradiance in wheat (Sofield *et al.*, 1977) or maize (Schoper *et al.*, 1982), but seed size increased. Higher individual seed growth rates in rice shortened the seed-filling period when there was no change in seed size (Kato, 1999).

The N supply to the plant during seed filling plays an important role in maintaining green leaf area during seed filling (Wolf *et al.*, 1988a; Banziger *et al.*,1994). Nitrogen stress accelerated leaf senescence (Boon-Long *et al.*, 1983; Hayati *et al.*, 1995) and shortened SFD in soybean without affecting SGR (Fig. 3.4). Increasing N fertilizer rates in the field lengthened the seed-filling period of soybean, bush bean (*Phaseolus vulgaris* L.) (Thies *et al.*, 1995) and sorghum (Kamoshita *et al.*, 1998), but wheat responded only when water was not limiting (Frederick and Camberato, 1995; Yang *et al.*, 2000). The yield response of maize to P and K fertilizer was related to an increase in SFD (Peaslee *et al.*, 1971).

The effect of increasing the supply of assimilate on SFD depends on, first, the effect of the treatment on senescence (how long is assimilate available?) and, second, the characteristics of the individual seed (can it continue to grow when assimilate is supplied for a longer period?). This interaction between supply and utilization determines whether changes in assimilate supply affect SFD. There are examples in the literature of both responses leading to a longer SFD in a variety of crops, but there are also examples where one or both of these responses failed, leading to no change in SFD. Accelerating senescence (water, nutrient or disease stress) will certainly shorten the seed-filling period and reduce seed size. Shortening the SFD and thereby reducing yield is probably more likely in the field than lengthening it and increasing yield.

Fruit and seed position

As mentioned previously in this chapter, anthesis or pollination may not occur at the same time for all flowers on a plant. The variation is relatively small in some crops (e.g. wheat) but it is large in other crops (e.g. soybean). Maturation of the developing seeds is, however, much more uniform, as it must be for efficient commercial production (Hay and Kirby, 1991). Variation in when seeds start to grow coupled with a relatively uniform maturation should lead to variation in SFD and, in fact, in many crops, seeds from the first flowers to pollinate have the longest SFD. In soybean, the SFD of seeds from flowers that were pollinated first (growth stage R1) was 36 days compared with a 27-day duration for seeds developing later (growth stage R4.5) (Egli *et al.*, 1987c). Gbikpi and Crookston (1981) reported similar results. Spaeth and Sinclair (1984) reported shorter SFD for seeds at upper nodes of soybean plants (probably from late-developing flowers) compared with seeds from lower nodes (probably from early flowers). Differences were also found when comparing the SFD of tip (short) and middle or basal seeds (long) of maize (Tollenaar and Daynard, 1978; Frey, 1981; Hanft et al., 1986), spikelet positions of wheat (basal seeds had a longer SFD) (Rawson and Evans, 1970; Simmons et al., 1982) and positions in panicle (top seeds had longer (rice) or shorter (sorghum) SFDs) (Jongkaeuwattana et al., 1993; Gambin and Borras, 2005).

Variation in SFD resulting from the flowering to maturation time of individual flowers probably accounts for some of the variation in seed size on a plant. The first flowers to develop often produce larger seeds than later-developing seeds (soybean, Egli et al., 1987d; wheat, Acreche and Slafer, 2006; maize, Tollenaar and Daynard, 1978), which is consistent with a longer SFD for earlier flowers. Seed size in soybean, however, was not consistently related to the time of flower development and it was hypothesized that variation in the length of the lag period of seed growth could account for some of this disconnect between the timing of development and seed size (Egli, 2012).

Miscellaneous factors

It's possible that plant hormones and photoperiod could affect SFD. Plant hormones affect leaf senescence (Lim et al., 2007) which could affect SFD; however, there is no evidence that they directly affect SFD. Suggestions that SFD in soybean is sensitive to photoperiod (Morandi et al., 1988; Han et al., 2006; Slafer et al., 2009) are not consistent with field observations. Although soybean is a photoperiod-sensitive species, the SFD was relatively stable across planting dates (Egli et al., 1987b), suggesting no photoperiod control.

Regulation of seed-fill duration

Little is known about the regulation of SFD, much less than is known about the regulation of SGR. The basic question that must be answered is – why does the seed stop growing? As with SGR, the answer to this question could reside in the ability of the mother plant to supply assimilate to the developing seed, or in the seed itself. The seed might stop growing because the plant no longer supplies it with C, N and other nutrients that drive seed dry matter accumulation or, alternatively, it could stop because some mechanism in the seed triggers processes leading to a cessation of dry matter accumulation and maturation when assimilate is still available. As with SGR, both mechanisms are involved in stopping seed growth.

Assimilate supply

Canopy photosynthesis and the redistribution of stored assimilate during seed filling provide the raw materials for seed growth. In most crop species, however, canopy photosynthesis begins an irreversible decline early in the seed-filling period (e.g. soybean, Wells et al., 1982, Christy and Porter, 1982; maize, Pearson et al., 1984; sunflower, Whitfield et al., 1989; wheat, Gent, 1995) and usually approaches zero as the seeds mature. Leaf senescence – 'the series of events concerned with cellular disassembly in the leaf and the mobilization of materials released during this process' (Thomas and Stoddart, 1980) – is responsible for this loss of photo-synthetic function that is normally characterized by the loss of chlorophyll, N and photosynthetic activity (Lim *et al.*, 2007). Consequently, the ability of the plant to produce assimilate for seed growth declines during seed filling, but it is not known whether canopy photosynthesis and the supply of assimilate reaches zero by physiological maturity. The data of Christy and Porter (1982) show that canopy photosynthesis of soybean was 10–20% of maximum rates at the end of seed filling, but they did not explain how they determined the end of seed filling. Pearson *et al.* (1984) also reported low levels of canopy photosynthesis at maturity in maize, but they also failed to define maturity. Most investigations of time trends of canopy photosynthesis did not report when physiological maturity occurred, so the data do not provide a definitive answer to the question of whether or how often canopy photosynthesis reaches zero before physiological maturity.

Premature termination of the assimilate supply by complete defoliation stopped seed growth much sooner than for the undefoliated controls. Seeds on defoliated plants went through a normal maturation sequence (i.e. colour change, loss of moisture, development of black layer, etc.) and the mature seeds were per-fectly normal in terms of shape, colour and ability to germinate, but they reached physiological maturity much sooner and were much smaller (Vieira *et al.*, 1992 – soybean; TeKrony and Hunter, 1995 – maize). Obviously, there will be no dry matter accumulation without a supply of raw materials from the mother plant, but whether this lack of assimilate is the normal trigger of the end of seed growth in the field is not clear.

Seeds will also mature normally when assimilate is still available, suggesting that a lack of raw materials for growth is not an absolute prerequisite for the cessation of growth. Seeds often mature when the plant is still photosynthetically active when source–sink ratios are altered in favour of the source (Munier-Jolian *et al.*, 1996), when stem reserves are not exhausted, and when the plant still has green organs (Banziger *et al.*, 1994, wheat; Jones and Simmons, 1983, maize). Physically restricted soybean seeds (i.e. increase in volume and dry weight was limited) matured, even though leaves were still photosynthetically active (i.e. chlorophyll levels and Rubisco activity were well above zero) (Crafts-Brandner, unpublished data, 1995; Egli *et al.*, 1987a). Soybean plants may retain green leaves at physiological maturity in max-imum yield environments (Purcell, 2008). These results clearly demonstrate that seed maturation can occur when assimilate is still available to the seed, i.e. mechan-isms intrinsic to the seed trigger maturation. Any answer to the question as to why the seed stops growing must accommodate cessation triggered by a lack of assimilate or by mechanisms in the seed that are independent of the supply of assimilates.

Seed water status

Seed water content and concentration are intimately related to dry matter accu-mulation during seed development in all crop species. Water content and seed volume always increase to a maximum before PM, while water concentration de-clines steadily, reaching a characteristic level at PM (Figs 2.2–2.4). These general patterns are the same for all crop species and in all environments; such consistency suggests that seed water relations may be the key to answering the question – why does the seed stop growing?

Since cell division is complete before the beginning of rapid dry matter accumulation, the increase in seed volume is a result of cell expansion driven by water uptake. This water movement into the seed results in the well documented increase in seed water content (mg seed^{-1}) and seed volume during the linear phase of seed growth (see Figs 2.2–2.4). Cell expansion, however, stops when the seed has its maximum volume and water content, which occurs before the end of dry matter accumulation. Dry matter accumulation continues and the water concentration declines until it reaches the level characteristic of physiological maturity when dry matter accumulation stops. The seed water concentration is intimately related to the stage of development in all species (Swank *et al.*, 1987; Calderini *et al.*, 2000; Borras and Westgate, 2006).

This description of seed growth has two main components – the period of cell expansion and water uptake that lasts until the seed reaches its maximum volume (and water content), followed by continued loss of water until the seed reaches the water concentration associated with PM. It is this water loss or 'desiccation' that may eventually trigger seed maturation. The key processes are those regulating water movement into the seed and the effect of seed water status on metabolic activity. Water movement into the cells in a seed is driven by the osmotic gradient across the cell wall (Lockhart, 1965), so assimilate availability could, therefore, regulate water movement into the cell and influence seed volume and potential seed size (Egli, 1990). Cell expansion could also be limited by physical restriction by fruit or seed structures (Egli, 1990). Following this model, a decrease in assimilate supply as a result of senescence, could stop water movement into the seeds (reduce the osmotic gradient), fixing the maximum seed volume after which desiccation continues until physiological maturity. On the other hand, physical restriction by fruit or seed structures could also limit cell expansion regardless of the osmotic gradient and trigger seed maturation even though adequate supplies of assimilate are available. Maximum seed volume is determined by the interplay of the movement of water into the seed and the maximum potential seed volume controlled by physical seed structures.

Evidence that tissue water status (water potential) regulates metabolic processes, in some cases at the gene level, supports this model (see Rodriquez-Stores and Black, 1994, for examples). Numerous authors (e.g. Walbot, 1978; Adams and Rinne, 1980) have discussed the evidence supporting a regulatory role for seed water status. The relatively constant water concentration at physiological maturity (the end of seed growth) within a species (Slafer *et al.*, 2009) (see Table 2.1) is consistent with the water status of the seed playing a regulatory role in seed development.

The continuation of seed growth beyond normal limits when cell expansion is allowed to continue, i.e. seed desiccation is delayed, is consistent with this model. Dry matter accumulation of soybean seeds in liquid culture with the testa removed continued much longer and the seeds were twice as large as seeds developing *in vivo* (Fig. 3.10, Egli, 1990). Seed volume continued to increase *in vitro* as water

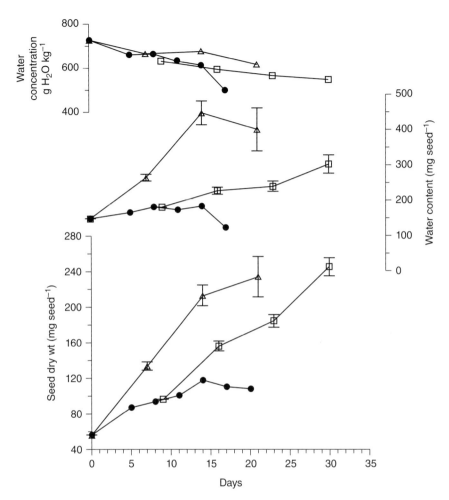

Fig. 3.10. The relationship between soybean cotyledon growth *in vitro* and seed growth *in vivo* (●). Cotyledons were placed in culture in a medium that contained 200 mM sucrose at 0 (△) and nine days (□). Cotyledon dry weight and water content were multiplied by two for comparison with seed growth rate *in vivo*. *In vivo* seeds reached physiological maturity on day 14. From Egli (1990).

moved into the seed and the water concentration remained above the critical level triggering seed maturation. If the testa remained intact and on the seed, water uptake and growth were greatly restricted. The addition of mannitol to decrease the osmotic potential of the media stopped water uptake and cell expansion and caused premature cessation of growth of soybean seeds when there were adequate supplies of C and N available (Egli, 1990).

Physically restricting the developing seed to limit the increase in seed volume reduced seed size in soybean (Egli *et al.*, 1987a; Miceli *et al.*, 1995), wheat (Grafius,

1978; Millet and Pinthus, 1984), barley (Grafius, 1978), oat (Grafius, 1978) and rice (Murata and Matsushima, 1975). In soybean, when two of three seeds in a pod were restricted, the unrestricted seed had a larger volume and final size (Fig. 3.11, Egli *et al.*, 1987a). When only part of a single seed was restricted, the unrestricted part was much larger than the restricted part, thus each seed or part of a seed responded to its ability to increase in volume. The size and shape of the space in which the wheat grain develops influences seed shape and size (Boshankian, 1918; Millet, 1986), seed size in rice is limited by the glumes (Murata and Matsushima, 1975; Jones *et al.*, 1979) and there is a good correlation between carpel size and seed size within and among legume species (Corner, 1951; Duncan *et al.*, 1978; Frank and Fehr, 1981; Fraser *et al.*, 1982a) and carpel or ovary size in spring barley (Scott *et al.*, 1983) and wheat (Calderini *et al.*, 1999). The role of the carpel in restricting seed expansion is illustrated vividly when soybean seeds occasionally expand with enough force to split the pod before the seeds reach maturity (Fig. 3.12). This phenomenon seems to occur in situations where pod size is reduced by stress (e.g. drought) and provides a vivid example of the role the pod plays in restricting seed expansion. Although we have observed this phenomena in several environments, it occurs so infrequently that experimental investigation is impossible.

This proposed model of seed maturation provides an answer to the question as to why the seed stops growing by focusing on the question – what stops the increase in seed volume (cell expansion)? The end of seed growth in this model is determined entirely by the seed's physiological environment, which is a function of the capacity of the plant to supply assimilate to the seed and morphological characteristics of the seed or fruit that limit the increase in seed volume. These characteristics can be influenced by the assimilate supply early in seed development. The model is not dependent upon an independent regulatory mechanism or 'clock' in the seed that stops growth when a certain time has passed.

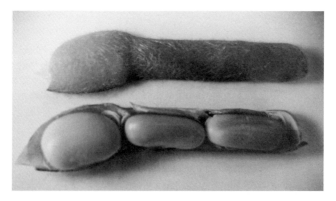

Fig. 3.11. Effect of physical restraint on seed size and shape in soybean. The two seeds on the right developed in the part of the pod where expansion was restricted by a plastic pod restriction device (PPRD) (Egli *et al.*, 1987a), placed on the pod at the beginning of seed growth. The seed on the left was not covered by the PPRD.

Fig. 3.12. Soybean pod that was split during seed development by the force of the developing seed. Seed growth stopped soon after the pod split, probably because the split pod allowed rapid seed desiccation.

The reported variation in seed moisture concentrations at physiological maturity among genotypes (maize, Carter and Poneleit, 1973) seed position in the inflorescence (sorghum, Gambin and Borras, 2007) and with premature maturation triggered by stress (Sala *et al.*, 2007b; Rondanini *et al.*, 2007) is not entirely consistent with this model, but the variation may simply reflect inaccurate estimates of the time of physiological maturity or the failure of bulk water concentration to accurately reflect the water status of the seed (Egli and TeKrony, 1997).

Although much of the work leading to the development of this model was with soybean (Egli, 1990), more recent work in other species (maize, Gambin *et al.*, 2007), and the general consistency of seed growth characteristics across species already noted, justify the application of this model to all grain crop species. Only additional research can determine whether this extension is appropriate.

The most important implication of this model is that both the size of fruit and seed structures, fixed at the beginning of the linear phase of seed growth, and the physiological environment of the seed can impose immutable restrictions on seed size and SFD by controlling the maximum seed volume and water content. Since fruit-seed structures are formed before rapid seed growth begins, the environment during the early development stages, as well as during seed filling, can play a role in determining final seed size. This simple model, which may have to be modified as more information becomes available, helps us understand why some seeds are large and others are small, why seed size and SFD are affected by the environment and why there is so much variation in the response of seed size to alterations in source–sink ratios, thereby improving our understanding of the yield production process.

Summary

The accumulation of dry weight by a seed and its ultimate final size is the product of a rate expressed for a specific time (i.e. SGR x SFD). Understanding the basis for the variation in SGR and SFD is the key to understanding, at a fundamental level, the variation in seed size. Seed growth rate and SFD are regulated by the seed and by the mother plant through its ability to supply assimilate to the developing seed. Genetic differences in SGR are regulated by the seed through the number of cells in the cotyledons or endosperm. The environment early in seed development can affect the number of cells, while the environment during seed filling can affect the metabolism in the seed directly or the supply of assimilate to the seed, all of which can influence SGR. Less is known about the regulation of SFD but the model developed in this chapter suggests that both the seed and the plant are involved.

The production of crop yield is often studied by dividing it into its components, seeds per unit area and seed size. The relationship of these components to yield can appear to be confusing and frequently contradictory. We can reduce this confusion by considering the fundamental processes underlying the yield components, i.e. the characteristics of seed growth – SGR and SFD. In the next chapter, we will use these characteristics to develop a basic understanding of the processes regulating the yield components; this understanding will clarify the relationships between seed number, seed size and yield.

Yield Components – Regulation by the Seed

4

Dividing yield into its components is essential to understanding the processes involved in the production of yield. The concept of yield, the weight of seeds at maturity, is a contrivance of humans; the plant does not produce yield, it produces flowers and then seeds that grow, accumulating complex carbohydrates, protein and oil, until they reach maturity. It is necessary to focus our attention on flowers and seeds to evaluate yield production at a physiological level. Focusing on the end product, yield, will not help us understand the process.

One of our objectives in this book is to use our knowledge of seed growth characteristics, developed in Chapters 2 and 3, to understand the role of the individual seed in the production of yield. Yield – the weight of seeds harvested from a unit area when the crop is mature – becomes a defined, measurable quantity only at the end of the crop's growth cycle, so it is difficult to relate processes occurring during earlier growth stages to the final yield. The key to making this connection is to focus on yield components and use characteristics of seed growth to understand their regulation and involvement in the yield production process. If we understand the regulation of the yield components, we understand the yield production process. We will find that involving seed growth characteristics will lead to a more profound understanding of how yield is produced and how it responds to plant characteristics and environmental conditions. For example, one common yield component is seed size (weight per seed), which may or may not be related to yield. Understanding the relationship between seed growth rate (SGR), seed-fill duration (SFD) and seed size will clarify this apparent ambiguity.

Yield components were used as early as the 1920s to analyse wheat yield responses to changes in plant population (Engledow and Wadham, 1923), but their popularity has varied in the ensuing century between periods of intense interest by plant breeders and crop physiologists, and benign neglect, when the focus was only on yield. Yield components are probably not as useful as the early practitioners hoped, but they cannot be ignored in any serious dissection of the yield production process.

Yield Components – Seeds per Unit Area and Seed Size

Historical use and misuse

The complexity of yield, apparent from the beginning of the discipline of crop physiology, may have stimulated interest in dividing yield into its components because the study of complicated systems is easier if they are divided into components and the components are studied separately (Charles-Edwards *et al.*, 1986, pp. 1–4). Plant breeders were also interested in yield components as they looked for more efficient breeding systems in their quest for higher yield. Perhaps selection for the components of yield would be more effective, leading to more rapid progress, than direct selection for yield. Unfortunately, focusing on yield components did not always improve our understanding of yield or increase the efficiency of breeding for yield.

The yield component approach by plant breeders was not successful, in part because many breeders encountered 'yield component compensation', where selection for one component was successful but other components adjusted so there was no change in yield.

A classic case of compensation occurs, for example, when selection for large seeds increases seed size but seed number decreases to maintain a constant yield. Such responses have been reported for common bean (White and Izquierdo, 1991) and soybean (Hartwig and Edwards, 1970) and probably exist for all crops. The phenomenon of yield component compensation also discouraged crop physiologists from using the yield component approach and subsequently many researchers focused entirely on studying yield.

The statistical basis of many yield component investigations also limited the usefulness of the component approach. Plant breeders and crop physiologists collected data on yield components in breeding populations, or collections of cultivars, and used statistical techniques, including correlation and path coefficient or factor analysis, to search for relationships between the components and yield. The usefulness of this approach was limited by the dependence of the results on the variation in the population under study, the intercorrelation of the components (i.e. yield component compensation), and a lack of consideration of the fundamental physiological processes regulating the individual components. Without consideration of physiological processes, many of the relationships defined by statistical analysis were not useful or mechanistically related to yield. Purely statistical associations among yield components contribute no more to our understanding of the yield production process than the correlation, often quoted in statistics classes, between stork numbers and human population size contributed to our understanding of human population dynamics.

The yield component approach was also hindered by the tendency of some researchers to create too many components. Dividing yield into a long list of components increases the complexity of the system instead of decreasing it, one of the original goals of yield component analysis, and increases the chances of defining components that are not particularly useful or meaningful. A long list of components also increases the chances of being misled by statistically significant relationships with no physiological basis.

Yield components are sometimes defined on a per-plant basis instead of a community or land-area basis (i.e. pods per plant instead of pods per unit area), which may create confusion for components that are extremely sensitive to plant population. A typical yield component equation for a grain legume, such as soybean (Equation 4.1), uses plant population (plants per unit area) to convert components per plant to an area or community basis.

$$\text{Yield}(\text{weight/area}) = (\text{plants/area})(\text{pods/plant})(\text{seeds/pod})(\text{weight/seed}) \quad (4.1)$$

Pods per plant (and seeds per plant, since there is usually little variation in seeds per pod) of soybean and other grain legumes varies inversely with plant population over a wide range of populations, so that pods per unit area and yield remains constant. Variation in pods per plant in this example could be solely a function of plant population and have no relationship to yield. Investigations using pods per plant or seeds per plant can be misleading and counter-productive unless population is carefully held constant for all comparisons, a rare occurrence for many crops in field experiments where population is not a treatment.

Yield components are frequently measured on individual plants selected from the community which can result in biased estimates if the plants are not representative of the population. Yield calculated from yield components estimated from several 'representative' plants may be much higher than yield measured traditionally by harvesting all plants from a specified area. For example, in a soybean population study, yield calculated from yield components determined on five plants per plot was 19–61% higher than yield estimated by harvesting the entire plot (Dominguez and Hume, 1978). Obviously, the plants harvested for the determination of yield components were not representative of the population, probably because small plants were not included in the sample. One cannot expect the yield components from a biased sample to accurately represent the response to treatments. Representative samples can be obtained by including all plants in a given area or in a specified length of row. The yield components from an area sample can be expressed on a per-plant basis using the number of plants in the sample, but the area-sampling technique will ensure that the yield components represent the plant community.

A more realistic equation for grain legumes that eliminates population-sensitive components can be created by combining components in Equation 4.1 to put the focus on the plant community (Equation 4.2), the unit that actually produces yield.

$$\text{Yield}(\text{weight/area}) = (\text{pods/area})(\text{seeds/pod})(\text{weight/seed}) \quad (4.2)$$

Combining pods per area and seeds per pod reduces Equation 4.2 to Equation 4.3.

$$\text{Yield}(\text{weight/area}) = (\text{seeds/area})(\text{weight/seed}) \quad (4.3)$$

This simple equation contains the primary components (seed number and seed size (weight per seed)) that determine yield, and it applies to all grain crop species, regardless of their growth habit. We will focus on this equation for the rest of this book. Equation 4.3 puts the focus where it should be, on the two primary components of yield and it avoids potentially confusing relationships created by including too many components. Each of the components in Equation 4.3, however, can be further divided to explore the basis of variation of that component after the importance of the component is documented. For example, seeds per unit area in a grain legume is a function of nodes per unit area, flowers and pods per node, flower and pod abortion, and seeds per pod. If seed number is related to yield, asking whether flowers or nodes account for the variation in seeds is a more direct, meaningful and testable question than asking whether, for example, flowers per node is related to yield. Creating additional components may increase the complexity of the system, but it is useful when the components are based on known relationships (e.g. flowers produce pods that contain seeds) and they do not represent a blind statistical search for relationships. I believe that creating additional components is consistent with the use of yield component analysis to simplify the system, making it easier to understand and fostering the development of mechanistic descriptions of the production of yield.

Equation 4.3 can be applied to all grain crop species, regardless of how their seeds are produced, which emphasizes the consistency of the yield production process across species. Even though seeds of grain crops are borne in many structures (pod, rachis, spike, compact ear), distributed over the plant or concentrated in one location, and vary greatly in size and/or composition, yield is always a function of the total number of seeds (seeds/area) and weight per seed (seed size). Equations containing more components (e.g. Equation 4.1) are species-specific, making it more difficult to develop general mechanisms describing the production of yield.

There have been times when yield component analysis fell into disrepute with many crop physiologists, but I feel that this occurred because of misuse and unrealistic expectations, and does not represent a fundamental flaw in the concept. Yield components do make a complex system simpler and easier to understand. But more importantly, we cannot hope to understand the mechanisms underlying the production of yield by only considering yield itself – we must consider the components of yield. After all, the plant doesn't produce 'yield', it produces flowers and then seeds which grow to maturity and are harvested as yield. Without involving yield components, we are unlikely to progress beyond the experimentalist approach of applying treatments and observing yield response, without proposing mechanisms. For example, it seems much easier to hypothesize and experimentally test mechanisms that account for the relationship between canopy photosynthesis and seed number than for relationships between canopy photosynthesis and yield. Focusing on yield provides no guidance on whether photosynthesis rate or duration is more important, or whether there is a critical plant growth stage when it is more important. These issues can be dealt with directly when yield components are included in the analysis.

Evaluation of yield components will often provide some insight into how the environment affected yield. If seed number is reduced, stress could have occurred during vegetative growth (possibly reducing radiation interception) or, more likely, during flowering and seed set. Seed size generally responds to variation in environmental conditions after seed number is fixed, so variation in seed size usually reflects environmental conditions during seed filling.

Studying yield components puts the focus on simpler systems that are easier to investigate and understand. Every component can generally be subdivided into more components, which should drive our understanding to ever lower, more definitive and narrower levels of organization. This top-down approach, i.e. starting at the level of the community and working down, will ensure that the knowledge gained at every level will provide useful information about the community. The weakness of the reductionist approach of many plant physiologists is that it starts at lower levels of organization without any clear picture of how the process chosen for study relates to higher levels of organization. Frequently there is no relationship. A top-down approach will eliminate this problem.

Yield components also contribute to the construction of mechanistic crop simulation models that realistically portray crop growth. Crop models without yield components frequently calculate yield as an empirical fraction of the total biomass, reducing yield physiology to the study of canopy photosynthesis and other factors associated with primary productivity; surely not a very rich, detailed evaluation of the yield production process. To model seed yield mechanistically, it is necessary to approach yield from its components and include mechanisms responsible for the regulation of these components. Including the primary yield components may increase the capacity of the model to respond to environmental fluctuations and thereby make the model more portable. The effect of environmental stress on yield is greatly dependent upon when in the plant's life cycle they occur, but it is hard to capture these effects in a model that calculates yield from total biomass and harvest index. Including yield components forces the model (or modeller) to relate to the growth stages when each yield component is sensitive to the environment.

Yield is a direct function of the number of seeds per unit area and the weight of the individual seed (seed size). Providing a mechanistic understanding of the production of yield requires consideration of these yield components; without such a consideration, our knowledge of yield will depend on empirical relationships.

Yield components and plant development

Yield is the final product of many environmentally sensitive morphological and physiological processes integrated throughout the 100 or more days of the life cycle of the crop. Integration of these processes over time creates many interactions between stage of development and the environment, greatly increasing the complexity of the yield production process. Yield components make it easier to understand these interactions, because each component is associated with a specific stage of plant development. Focusing on yield components helps define the production of yield as a sequential function of the stage of crop development (i.e. time).

Grain crops bear their seeds in a variety of configurations. The modern maize hybrid produces all of its seeds on a compact ear near the middle of the main stem. There are no ear-bearing branches or tillers, although prolific hybrids may produce more than one ear per plant. Wheat, barley and rye produce all of their seeds in a compact spike located at the top of the stem. A single plant may produce many tillers and each tiller may produce a seed-bearing spike. Rice and oat produce a panicle located at the top of the main stem and tillers. Sunflower produces all of its seeds in a compact head (capitulum) at the top of the main stem. All of these crops concentrate their seeds in a single location on the plant. Soybean and the other grain legumes (bean, pea, chickpea, broadbean, etc.) are quite different, with seeds borne in fruits (pods) that develop from flowers produced at nodes on the main stem and branches. In these species, seeds are distributed over the entire plant. Groundnut seeds are produced in the soil at the end of a peg or gynophore, which arises from a node on a lateral branch and produces only a single fruit. The relatively even distribution of fruits over the plant in these legumes results in only a small portion of the total seeds located at each node and the average distance from the source to the sink may be greatly reduced. In spite of this variation in seed-bearing characteristics, all grain crop species follow a similar sequential production of the principle components of yield.

Adams (1967), in his work with yield components, emphasized the sequential production of the individual components, and then Murata (1969) carried the sequential concept further by dividing the yield production process into three stages:

1. Formation of organs for nutrient absorption and photosynthesis.
2. Formation of flower organs and the yield container.
3. Production, accumulation and translocation of yield contents.

Murata's Stage One

Murata's first stage represents vegetative growth, when the plant produces the leaf area and roots that provide for and sustain canopy photosynthesis. The leaf area produced during this phase is a key component of crop productivity, because maximum canopy photosynthesis per unit ground area or crop growth rate occurs only when the leaf area is adequate to intercept most ($\geq 95\%$) of the incident solar radiation. Environmental conditions during vegetative growth are often thought to be unrelated to yield, as long as the crop reaches at least 95% solar radiation interception before the beginning of reproductive growth. Modern grain crop production systems have evolved to achieve maximum interception in most environments. Stress during vegetative growth can, however, reduce carbon capture during reproductive growth if leaf area and radiation interception are reduced below critical levels. Crop physiologists and agronomists generally agree that interception of solar radiation must reach a maximum at the beginning of reproductive growth, or shortly thereafter, to maximize yield of all grain crops.

Murata's Stage Two

Murata's (1969) second stage represents flowering, pollination, and the initial stages of seed growth, and it is the period when the seed number component of yield is determined. There are two aspects of the determination of seed number – the number of flowers that are produced and the survival of those flowers to produce mature seeds. At the end of this period, by definition, seed number is fixed and will no longer respond to changes in the environment. The initial formation of the yield container actually begins early in vegetative growth with the initial development of primordia of the structures that will ultimately bear the seeds. Maize produces ear initials at every node below the top ear-bearing node during vegetative development (Kiesselbach, 1949). Spikelet initiation in wheat begins very early in vegetative growth (Slafer *et al.*, 2009), but many flower-bearing nodes of soybean, and probably other grain legumes, form after the first flowers open (growth stage R1 in soybean). The exact beginning of Stage Two is crop-specific and identifying it precisely is difficult.

The processes leading to the production of a single seed or fruiting structure encompasses an orderly progression from the development of the initial primordium through the production of the flower, pollination and seed development. The process is not as orderly on a whole plant, because of variation in the timing of development of fruiting structures among locations on the plant. The magnitude of this variation is very species-specific; for example, the first flowers on a soybean plant (and probably other grain legumes) have pollinated and seed development is processing before the nodes that will produce the last flowers have appeared. In contrast, flower production will be more compact in time in crops where the flowers are borne on a single seed-bearing structure at the top of the main stem (wheat, rice) or a single ear located at a node in the middle of the plant (maize).

Environmental and cultural conditions in early vegetative growth can affect the morphogenesis of the seed-bearing structures, which could potentially affect seed number. These effects are not widely documented, but the data suggests they are quite variable among species and environments. For example, increasing plant density decreased florets per spike in wheat and the effect was apparent 20–24 days before anthesis (Yu *et al.*, 1988). There is some evidence that environmental conditions can influence flowers per node in soybean (Jiang and Egli, 1993; Egli and Bruening, 2006b). Ear size (kernel rows per ear and florets per row) of maize, however, was relatively insensitive to management and environmental conditions in some experiments (Siemer, 1964; Lemcoff and Loomis, 1986; Uhart and Andrade, 1995a) but N and defoliation stress (Jacobs and Pearson, 1991) and shade (Hashemi-Dezfouli and Herbert, 1992) reduced spikelets per ear in other experiments. Effects early in vegetative growth would be much less important in species, such as soybean, where many flower-bearing nodes are formed after flowering begins.

Species diversity and the potential influence of events in early vegetative growth on seed number make it difficult to develop a meaningful definition of the beginning of Murata's (1969) Stage Two that can be applied to all grain crops.

It is not always clear how sensitive these initial events are to environmental conditions and how closely they relate to final seed number, suggesting that it may not be necessary to include these early events in a useful description of Stage Two.

Grain crops usually have the capacity to produce a yield container that is much larger than needed. The production of flowers is often much greater than the number of seeds or fruits that survive to maturity, thus, potential seed number (the number of seeds if all flowers produce seeds that survive to maturity) is greater than final seed number. Potential seed number increases as the plant produces more nodes, tillers, flower primordia, and flowers, and it declines as reproductive structures do not continue development (abort), e.g. tillers fail to produce spikes, flower primordia fail to develop into flowers, flowers are not pollinated, and pollinated flowers or developing fruits abort. Consequently, the determination of seed number in commercial production systems is generally a process of reducing this 'excess' capacity to the number of seeds that the vegetative plant can support. This downward adjustment occurs during the critical period for seed number determination and this period can be practically taken as Murata's (1969) Stage Two, i.e. the period when seed number is determined. This definition of Stage Two doesn't include the initial processes leading to flower production, but, while those early events may affect potential seed number, they are not directly involved in the downward adjustment to final seed number. Defining Stage Two in this manner results in a stage that can be defined with reasonable accuracy in all crops, and relates directly to the processes involved in the determination of seed number. This definition does not relate directly to crops that are sink limited, i.e. every flower produces a seed that survives to maturity and there is no downward adjustment. Such limitations are probably rare in most modern agricultural systems.

This definition of Stage Two is consistent with the common description of the critical period used in the literature. Tollenaar *et al.* (2000), using data from Classen and Shaw (1970), demonstrated that kernel number was sensitive to drought stress from ~15 days before to 20 days after 75% silking, which is consistent with other reports for this crop (Andrade *et al.*, 2000, 7 days before to 14 days after silking; Echarte *et al.*, 2000, 10 days before to 15 days after silking). The critical period for seed number determination for soybean is probably from growth stage R1 (initial bloom) to between growth stages R5 (beginning seed fill) and growth stage R6. Board and Tan (1995) suggested that the end of the critical period occurred 10 to 12 days after growth stage R5, while Egli (2010) reported that 60% shade applied at the beginning of growth stage R6 reduced seed number. Seed number was no longer sensitive to the assimilate supply after that growth stage, some 45 to 55 days after growth stage R1. Small pod production (~ 10 mm long) often continued past growth stage R5 (Egli and Bruening, 2006a; Egli, 2013), suggesting that those plants had some potential to respond to an increase in photosynthesis after growth stage R5, assuming those pods were destined to abort without a change in assimilate availability. Shade treatments reduced kernel number of wheat between 30 and 40 days before and 10 to 15 days after anthesis (Fischer, 1975). More recently, the duration of the spike elongation phase, roughly 20 days before to 10 days after anthesis, was related to seed number (Fischer, 2011). The critical period has also been defined for sunflower (Cantagallo *et al.*, 1997), chickpea, (Lake and Sadras, 2014) and probably for many other crops.

Fischer and Laing (1976) thinned wheat plots to increase photosynthesis and found that thinning after anthesis had no effect on seed number, but thinning before anthesis caused large increases in seed number (Fig.4.1). Comparing these results to the effects of shade (Fischer, 1975) suggests that the timing of the end of the critical period may depend on whether it is based on stress that reduces seed number or improved environmental conditions that increase seed number. It is likely that the capacity to increase seed number will be lost before the capacity to

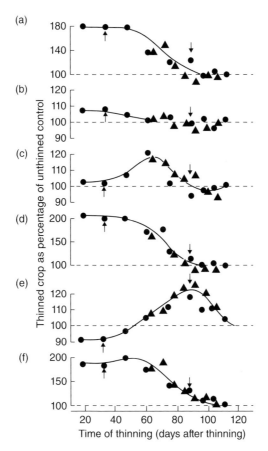

Fig. 4.1. The response of wheat yield and its components to thinning at different times after planting to increase solar radiation and photosynthesis per plant. Yield and yield components are expressed per metre of row, (a) spikes m^{-1}, (b) spikelets spike^{-1}, (c) seeds spikelet^{-1}, (d) grains m^{-1}, (e) weight seed^{-1}, and (f) total seed weight m^{-1}. Up-directed arrows = mean date of floral initiation, down-directed arrows = mean date of 50% anthesis. ● and ▲ represent different experiments. From Fischer and Laing (1976).

decrease seed number. Increasing seed number may depend upon the availability of flowers to develop into fruits and seeds, while fruits or seeds that are in the early stages of development can abort decreasing seed number. Environmental conditions and management practices during early vegetative growth before Stage Two can influence flower production and potential seed number, but they will have no direct effect on the downward adjustment to the final seed number that is fixed at the end of Stage Two. The early environment could, however, have an indirect effect on seed number if leaf area and solar radiation interception during Stage Two are reduced.

Clearly there is a critical period in the development of all grain crops when seed number is determined, and after that seed number will no longer respond to changes in environmental conditions. Murata (1969) defined Stage Two to include flower production, but it is difficult to practically apply this definition because flower formation can begin early in vegetative growth. Restricting Stage Two to the period when potential seed number is reduced to the final level is much more manageable. We will use this definition in the rest of our discussion of yield components and the yield production process. The fact that the exact beginning and end of the period are poorly defined in many crop species does not detract from the usefulness of this concept. Variation in the length of this period among crop species makes one wonder whether there is a relationship between length and seed number or the stability of seed number. This question will be explored in detail in Chapter 5.

Murata's Stage Three

Stage Three is the seed-filling period, when the seeds accumulate oil, starch and protein, i.e. when yield is actually produced. Interestingly, at the beginning of Stage Three no yield has been produced. All yield is produced during Stage Three, making it a critical period; all prior events are only preliminary preparations for the main event. Stage Three ends when the seed reaches its maximum size (weight per seed) at physiological maturity, so seed size is an indicator of how well the yield container is filled (how much of the potential yield is realized). Yield is therefore determined by the size of the yield container (seed number and potential seed size) and how well it is filled. The sink limits yield if the yield container cannot hold all of the contents available during Stage Three.

Summary

Murata's (1969) three stages of crop growth clearly describe the production of yield as a sequential process. First the plant grows vegetatively, producing leaf area for photosynthesis, then it flowers and sets seed, and finally it fills the seeds and matures. The last two stages relate directly to the two terms in our yield component equation (Equation 4.3), so the yield component equation also emphasizes the sequential nature of yield production. These stages occur in all grain crops, although the length and timing of the individual phases will vary among species, depending on plant characteristics, growth habit and morphological development.

Murata's (1969) stages one and two are not mutually exclusive in all crops. For example, vegetative growth in soybean continues during flowering and podset, so most of Stage Two occurs during Stage One. In contrast, vegetative growth stops at anthesis in other crops, such as wheat and maize, but even in these crops there is some overlap of Stages One and Two because assimilate stress just before anthesis can reduce seed number. Stage One probably ends before the end of Stage Two in most species. Stage One is probably complete when Stage Three begins in most modern crop cultivars but there is evidence that vegetative growth continued during seed filling in old cultivars of some species (Duncan *et al.*, 1978; Gay *et al.*, 1980) and probably in wild progenitors of many crops. A separation of vegetative (Stage One) and reproductive growth (Stages Two and Three) may contribute to maximizing partitioning of assimilate to seeds which is often assumed to be a necessary condition for maximum yield. Issues of partitioning are complex and not very well understood in most crops; this issue will be discussed in detail in Chapter 5.

Yield components and yield

In spite of the positive aspects of yield components emphasized in the previous discussion, the relationship between the simplified yield components – seed number and seed size (weight/seed) – and yield can be somewhat convoluted, complicating the use of yield components to understand the yield production process. Much of this complexity is related to the source of variation in the components – genetic or environmental. We can illustrate this complexity by looking at the relationship between seed number and yield.

Environmentally induced variation in yield is usually closely associated with seeds per unit area. In Fig. 4.2 and 4.3, most of the variation in yield of a single soybean or wheat cultivar across locations and years was associated with variation in seeds m^{-2}. Similar relationships have been shown for maize (Fig.4.4), sunflower (Cantagallo *et al.*, 1997), and field pea (Poggio *et al.*, 2005), and surely exist for all grain crop species.

There are, however, other situations where seed number and yield are not related. There can be substantial differences in seed number among cultivars within a species that are not related to yield, as shown in Table 4.1 for soybean. Species differences in seed number are usually not related to yield (Table 4.2). Average yield of sorghum was less than half the yield of maize but seed number of sorghum was four times that of maize. Wheat and soybean had somewhat similar yields, but a roughly fivefold difference in seed number per unit area. Variation in seed number explains much of the yield variation among environments, but it is not always related to the variation among cultivars or species (i.e. genetic variation).

The relationship between yield and seed size is probably more complex than yield and seed number. The substantial environmentally induced variation in seed size in the data sets represented in Fig. 4.2 and 4.3 was not significantly related to yield. Reducing photosynthesis with shade reduced yield in both experiments in

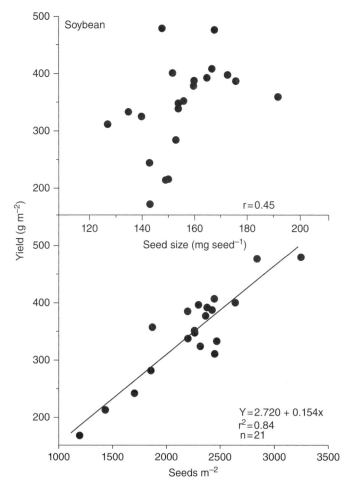

Fig. 4.2. The relationship between soybean yield and the yield components seeds m^{-2} and seed size. Cultivar Iroquois (maturity group III), 21 locations in 1996. Unpublished data from the Uniform Soybean Test – Northern region.

Table 4.3, but seed size was smaller only when the treatment was applied during seed filling (~ growth stage R6; 1993–1995). Applying the treatment was during the entire reproductive growth period (growth stage R1 to maturity, 1989–1990) reduced yield by 54%, but there was no change in seed size because all of the reduction was a result of fewer seeds. The environment affects seed size only during seed filling, after seed number is fixed. Since seed number accounts for most of the adjustment to the environment, environmental effects on seed size are much less important, but they can contribute to yield variation. Opportunities for stress during Stage Three to reduce seed size and yield are probably more common than for favourable environments to increase seed size and yield.

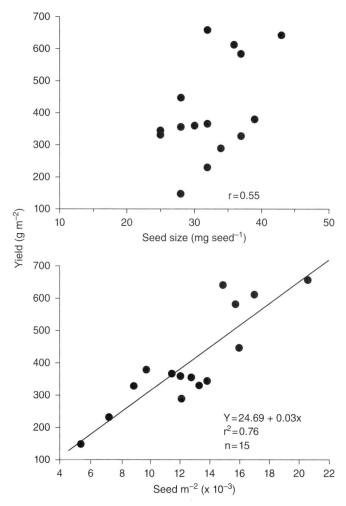

Fig. 4.3. The relationship between wheat yield and the yield components, seeds m⁻² and seed size. Cultivar Cardinal, 15 location-years. Unpublished data from the 1990/91 to 1995/96 Uniform Eastern Soft-Red Winter Wheat Nursery Program.

Cultivar differences in seed size within a species show the same inconsistent relationship to yield. These genetic differences are sometimes related to yield, as illustrated by the findings of Gay *et al.* (1980) that the then relatively new soybean cultivar (Williams) out-yielded an older cultivar (Lincoln) by 34% with no difference in seed number. Williams had larger seeds than Lincoln, which accounted for all of the difference in yield. There are, however, other comparisons where cultivar differences in seed size were not related to yield (Table 4.1). Comparisons among species show little relationship between seed size and yield. Rice is tied with sorghum for the smallest seed of the species in Table 4.2, but it has the

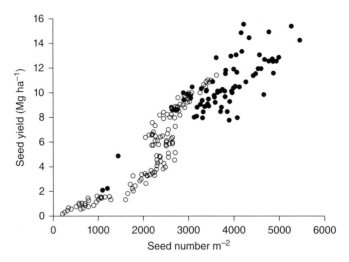

Fig. 4.4. The relationship between maize yield and seeds m⁻². Open symbols
from Chapman and Edmeades (1999) and closed symbols from Otegui (1995).
From Borras *et al.* (2004).

Table 4.1. Yield and yield components of two soybean cultivars
with differences in seed size, 1989–1990 (Egli, 1993).

Cultivar	Yield (g m⁻²)	Seed number (no. m⁻²)	Seed size (mg seed⁻¹)
Harper	337	1668	200
Essex	330	2156	152
	NS	*	*

*Significant at $\alpha = 0.05$, NS = not significant.

Table 4.2. Species differences in seed size, seeds per unit area and yield.

Species	Approximate seed size[1] (mg seed⁻¹)	Average yield[2] (g m⁻²)	Seeds per unit area[5] (no. m⁻²)
Rice	28	849	30321
Wheat[3]	41	296	7220
Sorghum	28	424	15143
Maize	302	1039	3340
Soybean	202	312	1544
Bean[4]	345	201	583

[1]From Table 3.1.
[2]Average US yields for 2013, 2014 and 2015. Data from National Agriculture Statistics
Service (www.nass.usda.gov).
[3]Winter wheat.
[4]*Phaseolus vulgarius* L.
[5]Seed number = yield/seed size.

Table 4.3. Shade stress, yield and yield components in soybean.

Cultivar/treatment		Yield (g m^{-2})	Seed size (mg seed^{-1})
1989–1990[1]			
Hardin	Control	294	142
	Shade[2]	138*	142 NS
Essex	Control	330	152
	Shade	152*	148 NS
1993–1995[3]			
Hardin	Control	345	151
	Shade[4]	261*	129*
Essex	Control	362	139
	Shade	286*	123*

*Shade treatments significantly different from control ($\propto$ = 0.05); NS = not significant.
[1]Adapted from Egli (1993).
[2]63% shade applied from R1 to maturity.
[3]Egli (1997).
[4]63% shade applied from early seed filling (approximately growth stage R6) to maturity.

second-highest yield. Bean had the largest seed and the lowest yield. One could argue that the average yields in Table 4.2 misrepresent the yield potential of each species, but such disparities were maintained in record yield environments, where, for example, sorghum yield was nearly equal to maize (Evans, 1993, pp. 288–289), but sorghum seed is smaller by a factor of 10. Clearly, large seeds do not always carry the connotation of high yield.

Even with the simplified yield component equation (Equation 4.3), the relationship between the individual yield components and yield is confusing because variation in either component may or may not be related to yield. It is easy to understand why the investigation of yield components has often been discouraging. The challenge is to develop mechanistic relationships describing the variation in yield components that provide an explanation for these diverse associations. Relationships based on the growth characteristics of the individual seed will provide a complete understanding of yield components and yield component compensation.

Determination of Seed Number

Components of seed number

The production of reproductive structures that ultimately bear the flowers and seeds during stage two follow a unique pattern for each crop species providing numerous opportunities for the environment to affect the individual components of seed number. Since the determination of the components that determine potential seed number is species specific, we will consider the process in some detail for three important crops with diverse growth habits, soybean, wheat, and maize.

Soybean

The components of seeds per unit area in soybean are given by Equation 4.4.

$$\text{Seed/area} = (\text{plants/area})(\text{nodes/plant})(\text{pods/node})(\text{seed/pod}) \quad (4.4)$$

Pods/node is determined by flowers/node and the proportion of flowers that produce mature pods expressed as a percent (SET, where SET = 1 - proportion of flowers or pods that abort) (Equation 4.5).

$$\text{Pods/node} = (\text{flowers/node})(\text{SET}) \quad (4.5)$$

The factor SET is less than 1.0 because flowers are not pollinated (an unlikely occurrence in soybean, Abernathy *et al.*, 1977) or because pollinated flowers fail to develop into mature pods (i.e. either flower or pod abortion occurs), a much more likely occurrence. Pods per node is thus a function of the number of flowers per node and SET.
Substituting Equation 4.5 into 4.4 produces Equation 4.6.

$$\text{Seeds/area} = (\text{plants/area})(\text{nodes/plant})(\text{flowers/node})(\text{SET})(\text{seeds/pod}) \ (4.6)$$

Plant population (plants/area) is determined by the seeding rate, selected by the producer, and the proportion of the seeds that emerge and produce a plant, a function of the quality of the planting seed and the seedbed environment. Nodes/plant is inversely related to population; increasing population results in fewer nodes per plant with most of the adjustment occurring on branches. Plant population can be eliminated by combining plants/area and nodes/plant in Equation 4.6 to produce Equation 4.7.

$$\text{Seeds/area} = (\text{nodes/area})(\text{flowers/node})(\text{SET})(\text{seeds/pod}) \quad (4.7)$$

All of the terms in Equation 4.7 are sensitive to the environment and can contribute to variation in seed number. For example, Hardman and Brun (1971) increased nodes by growing soybean plants at above normal CO_2 levels during vegetative growth, while shading plants during vegetative growth reduced the number of nodes on the main stem and branches (Jiang, 1993). Delayed plantings reduced nodes per plant (Egli *et al.*, 1985b; Bastidas *et al.*, 2008), but cultivars with long vegetative growth periods (i.e. late-maturing cultivars) produced more nodes per plant (Egli *et al.*, 1985b; Egli, 1993) and more per unit area (Egli, 2013) than early maturing cultivars with short vegetative periods. These environmental effects on nodes per plant would be reflected in nodes per unit area. Planting patterns may also affect nodes per unit area (Egli, 1994b; Ball *et al.*, 2001; Egli, 2013). Determinate cultivars that stop main stem growth after flowering begins may produce more nodes, owing to increased branching, than indeterminate cultivars of similar maturity that continue main stem growth until the end of the flowering period (Egli, 1994b).

Less is known about the factors affecting flowers per node, but there is evidence of environmental and genetic effects. Flowers per node varies by position of the node on the main stem (Brevedan *et al.*, 1978; Jiang and Egli, 1993). Reducing photosynthesis during flowering and podset by shade (Jiang and Egli, 1993) or defoliation (Bruening and Egli, 1999) reduced the number of flowers per node, suggesting that the variation could be related to assimilate availability at each node (Egli, 2015a). Genotypic differences in flowers per node were reported by van Schaik and Probst (1958), Jiang and Egli (1993) and Bruening and Egli (1999). In the latter two reports, small-seeded cultivars had more flowers per node than large-seeded cultivars.

A significant proportion of flowers and small pods do not survive until maturity in soybean, making SET (1 – abortion) an important part of the determination of seed number. Flowers and immature pods contribute to reproductive failure in an approximately 1:1 ratio (Heitholt *et al.*, 1986); the abortion of full-size pods (pods that have reached their maximum length) is, however, very rare (Egli and Bruening, 2006b). There is no evidence in the literature that SET ever approaches one, i.e. abortion is zero. High rates of flower and pod abortion can occur in high-yield environments, for example, 50–70% flower plus pod abortion occurred in a non-stress environment when yields were relatively high (400 g m^{-2}) (Jiang and Egli, 1993), suggesting that flower or pod abortion is part of the normal growth of the soybean plant and not just a response to stress. Flower and pod abortion, however, is increased by stress (Mann and Jaworski, 1970; Neyshabouri and Hatfield, 1986; Heitholt *et al.*, 1986; Jiang and Egli, 1993), but stress can also reduce the number of flowers per plant (by reducing flowers per node or nodes per plant), so that seed number can be reduced without any effect on SET (Jiang and Egli, 1993). In fact, the substantial variation in seeds m^{-2} across environments (Fig. 4.2) is probably more closely associated with variation in nodes and flowers m^{-2} than it is with variation in SET (Jaing and Egli, 1993; Egli, 1994b). The relative importance of SET compared with nodes m^{-2} or flowers node^{-1} may be related to the timing of stress. Stress occurring throughout Stages One and Two would probably reduce nodes m^{-2} and/or flowers per node, while stress occurring only during Stage Two may be more likely to decrease SET below normal levels.

Soybean plants produce pods that contain from one to four seeds, but three-seeded fruits are probably most common. Individual seeds in a fruit can abort in response to stress, resulting in, for example, a fruit with three locules but only two seeds. Seed per fruit generally does not show much environmental variation, so it is usually not an important contributor to variation in seeds m^{-2}.

The sequential nature of the seed number determination process can be clearly seen in Equation 4.7. First the nodes are produced on the main stem and branches, then flowers develop and are pollinated and some survive to produce mature pods. The process, however, is more complicated than implied by Equation 4.7, because of the sequential production of nodes throughout Stage Two and the sequential production of flowers at each node. The production of flowers at an individual node can continue for 20 or more days (Egli and Bruening, 2006a) while it can be 40 days or more from the first flower to the last flower on a plant (Egli and Bruening, 2005; Egli, 2010). In the middle of Stage Two there may be fruits containing seeds in the linear phase of growth at the bottom of the plant, higher nodes that have just started flowering and nodes still being formed at the top of the plant. This complexity makes it difficult to describe the production of potential seed number as a simple sequential process and complicates evaluation of environmental effects. Other crop species are less complex and provide a better fit to a simple sequential model.

The mechanisms responsible for adjustment of some components of Equation 4.7 are still unknown. For example, little is known about the development of the flower primordia at each node, when they develop and when, if ever, the number of primordia reaches a maximum. The mechanisms responsible for environmentally induced changes in flower number per node are unknown. The general nature of the process, however, is clear: large numbers of reproductive structures are produced and then aborted to establish the final seed number.

Wheat

The components of seeds per unit area of wheat are given by Equation 4.8.

$$\text{Seeds/area} = (\text{plants/area})(\text{tillers/plant})(\text{spikes/tiller})$$
$$\times (\text{spikelets/spike})(\text{florets/spikelet})(\text{SET}) \qquad (4.8)$$

Tiller formation in wheat starts soon after emergence and is closely related to leaf emergence (Slafer et al., 2009). Tillers per plant is inversely related to plant population (Bremner, 1969; Fraser et al., 1982b), resulting in a relatively constant seeds per unit area for a range in plant populations (Puckride and Donald, 1967) over which yield remains constant. Tillers per plant or per unit area are also influenced by environmental conditions, including availability of N (Bremner, 1969; Fischer, 1975) and extra solar radiation (Fischer, 1975). Since tillers per plant is related to population, it is not a particularly useful component of seed number; tillers per unit area is much more meaningful. Some tillers fail to produce spikes (spikes per tiller is equal to or less than 1.0) and the number of spike-bearing tillers usually reaches a maximum and then declines to the final number at anthesis (Fischer, 1975; Slafer et al., 2009).

Spikelets per spike and florets per spikelet are also influenced by the environment and management practices (Slafer et al., 2009). Evans et al. (1975) summarized reports demonstrating that spikelets per spike was influenced by solar radiation levels, planting density, defoliation, and N fertilization. The inflorescence of wheat is determinate and once the terminal spikelet is formed, spikelets per spike can no longer increase and the maximum number of spikelets is fixed.

Equation 4.8 could be further simplified by considering only florets per spike instead of its components – spikelets per spike and florets per spikelet – leading to Equation 4.9.

$$\text{Seeds/area} = (\text{plants/area})(\text{tillers/plant})(\text{spikes/tiller})$$
$$\times (\text{florets/spike})(\text{SET}) \qquad (4.9)$$

Yu *et al.* (1988) and Slafer *et al.* (2009) demonstrated that florets per spike were affected by plant population in several environments. Production of floret primordia greatly exceed the number of florets at anthesis with declines of 60–70% from the maximum reported by Yu *et al.* (1988). The SET term in Equation 4.9, accounting for the failure of component florets to produce mature seeds, is usually very close to 1.0 (Yu *et al.*, 1988; Slafer *et al.*, 2009), much higher than it is in soybean.

The sequential nature of the development of the components of seed number in wheat is just as evident as it was in soybean. First, the main stem and tillers develop and produce spikes and the development of florets in the spike is followed by pollination, fertilization and the beginning of seed growth. The compact nature and pattern of development of the wheat inflorescence suggests that development occurs over a shorter time span than in soybean. Most of the downward adjustment in wheat occurs before anthesis, when the tillers fail to produce spikes and floret primordia fail to develop component flowers.

Maize

Equation 4.10 describes the components of seed number for maize.

$$\text{Seeds/area} = (\text{plants/area})(\text{ears/plant})(\text{rows/ear})(\text{spikelets/row})(\text{SET})$$
$$(4.10)$$

Domestication greatly reduced the capacity of the individual maize plant to adjust the number of seeds it produces. The wild progenitors of maize produced many small ears on each plant and they had the capacity to produce branches and tillers (Mangelsdorf *et al.*, 1967). Modern hybrids have the potential to produce an ear at every node below the uppermost ear (Kiesselbach, 1949), but when grown at recommended populations, they usually produce only one ear. Low populations (Thomison and Jordan, 1995) or a lack of stress (radiation, water or N) encourage formation of a second ear (Motto and Moll, 1983), but the level of prolificacy (tendency to produce more than one ear per plant) varies among hybrids (Harris *et al.*, 1976; Motto and Moll, 1983). Florets per ear is determined by the number of rows per ear and the number of spikelets per row. Each spikelet contains two florets but only one develops (Kiesselbach, 1949), making spikelets and florets per ear synonymous. The ear primordia develop very early during vegetative growth, but the maximum spikelets per row on the top ear is reached only a week or so before silking (Siemer, 1964). Rows per ear and spikelets per row are under genetic control (Duncan, 1975), but they can also be influenced by environmental

conditions during vegetative growth. The response, however, seems to be relatively modest and inconsistent. Low radiation (Uhart and Andrade, 1995a), N
stress (Lemcoff and Loomis (1986) and planting date or year (Seimer, 1964) had
no effect of on the number of rows per ear or the number of spikelets per row.
Jacobs and Pearson (1991), however, found that N and defoliation stress reduced
both components, while shade stress reduced spikelets per ear (Hashemi-Dezfouli
and Herbert, 1974).

Plant population is an important component of seeds per unit area in maize
because modern hybrids no longer have the flexibility to substantially increase
seeds per plant; consequently, increasing population (more ears per unit area) is
the only way to significantly increase the number of seeds per unit area (Egli,
2015b). Some modern US hybrids will produce tillers at low plant populations but
these tillers generally do not produce seeds that contribute to yield; however, there
are other cultivars and hybrids that produce part of their yield on tillers (Duncan,
1975; Goldsworthy and Colegrove, 1974).

The SET factor, representing the proportion of florets that produce seeds,
can be 1.0 or less. Two distinct processes account for SET being less than 1.0 in
maize. First, a lack of pollination or fertilization is a relatively frequent occurrence in maize. High-temperature stress can reduce pollen viability (Herrero and
Johnson, 1980), while drought stress reduces silk receptivity (Basseti and Westgate,
1993b) and causes asynchronous flowering, so that pollen is shed when no viable
silks are present. Asynchrony has also been associated with reductions in seed
number at high populations and under N stress (see Jacobs and Pearson, 1991
and the references therein). Regardless of the cause, poor pollination or fertilization can result in large reductions in seed number. Second, fertilized flowers may
abort and not produce mature seeds; in this situation, SET is analogous to SET
in wheat or soybean. Pollination and fertilization issues are much more important
in maize than in soybean or wheat, and can result in catastrophic reductions in
seed number.

The determination of seed number in maize shows the same general characteristics as soybean and wheat. Seed number develops sequentially and there are
several components responsible for adjusting the number. Modern maize hybrids,
however, have fewer mechanisms of adjustment than either soybean or wheat,
consequently, seed number per unit area is much more dependent upon plant
population. When maize is managed correctly and grown at an optimum population, potential seed number at silking is greater than final seed number (i.e. all
spikelets don't produce mature seeds), so the determination of seed number is a
downward adjustment to the final number.

Summary

We have developed equations describing the components of seed number in three
major crops: a grain legume – soybean; and two cereals – wheat and maize. All three
equations contain different terms, reflecting the morphological characteristics of

the individual species; however, the principles governing determination of seed number are generally the same for all three species and only the details differ. Although detailed information on the formation of the yield container is not available for all grain crop species, it seems likely the general patterns described for soybean, wheat and maize apply to all other species.

These equations for all three species describe the determination of seed number as sequential in time, with each of the components that ultimately produce a fruit or seed formed one after another during plant development. Plant development on a plant basis is, in reality, never that simple. The timing of development of fruits or seeds may differ by location on the plant or in the reproductive structure. At any time, the fruits or seeds on the plant may be in different stages of development. In spite of this potential complexity at the plant level, some components are formed and fixed before others, so each component could be exposed to and influenced by a unique environment during its development.

The variation in the timing of development on a plant may affect how the plant responds to short-term changes in the environment. Theoretically (Egli, 2015a), species with large amounts of variation in the timing of individual seed development, resulting in a longer critical period (Stage Two), e.g. soybean, should be more resilient than species with more compact development, e.g. maize. The longer critical period would make it less likely that stress would affect the entire period, but there is little data available to support this proposition. The early developing components (nodes, tillers, primordia of reproductive structures, etc.) would probably make most of the adjustment to season-long differences in productivity, while short-term fluctuations in the environment would affect those components developing at the time of the stress. The early-developing structures represent a coarse adjustment to the environment, while the later-developing structures represent a fine adjustment, borrowing the concept developed by Slafer *et al.* (2014). In both cases, seed number adjusts to the environment, but the components responsible differ.

The potential seed number of all three species usually seems to be larger than the actual seed number (assuming maize is grown at the appropriate population), so the determination of seed number is a downward adjustment during the critical period. It is clear that crop plants can respond to a favourable environment by producing more seed-bearing structures (branches, tillers, nodes, ears etc., although maize requires human intervention to increase populations, so a large seed number at maturity is not always a result of decreasing the downward adjustment. In fact, much of the variation in seed number depicted in Fig. 4.2–4.4 is more likely to be due to variation in potential seed number than in the downward adjustment process. The downward adjustment process may be more important in short-term stress situations.

Some of the components of seed number described in Equation 4.7, Equation 4.9 and Equation 4.10 may also be under genetic control. Examples of genetic control include cultivar differences in flowers per node in soybean (van Schaik and Probst, 1958; Jiang and Egli, 1993), tillering in wheat (Evans *et al.*, 1975) and ear number in maize (Harris *et al.*, 1976). Species (cultivars) that produce small seeds will produce more flowers than large-seeded species (cultivars) (Jiang and Egli, 1993). Many other examples can be found in the literature. Thus, the magnitude of potential seed number, final seed number and the excess capacity could be a function of the cultivar as well as the environment.

Seed number component equations could be written for other grain crops and they would probably be similar to Equation 4.7–Equation 4.10. Equations for other cereals (e.g. barley, oats and rye) would surely be similar to Equation 4.9 and the soybean equation (Equation 4.7) could no doubt be applied to other grain legumes, such as pea or common bean. Other species may require quite different components (e.g. sunflower or oil-seed rape), but all equations would encompass the concepts of sequential determination of several individual components and a potential seed number that is greater than seed number at maturity. The general similarity among species encompasses a lot of variation in detail; for example, a process (e.g. SET) that is important in one species may be relatively unimportant in another species (soybean vs wheat). The sequential development is clearer in some species than others and some crops may have more components than others. But even with this variation in detail, the general principles are the same. The similarities across species in the general principles governing the determination of potential seed number and seed number suggest that the same mechanism(s) may be operating in all species.

Environmental effects

Most of the environmentally induced variation in yield of all grain crops is related to variation in the number of seeds per unit area (Fig. 4.2–4.4). Seed number is the first yield component determined, so it represents the first opportunity for the crop to adjust its reproductive output to the productivity of the environment. The close association between environmental conditions and seed number is not, therefore, surprising. Seed size, which is determined after seed number has adjusted to the environment, is much less variable. The consistency of this response across all grain crops supports the suggestion that the mechanisms governing the determination of seed number are similar for all grain crops. The consistency of seed number–environment relationships can be illustrated by considering examples for soybean, wheat and maize.

Reducing canopy photosynthesis by decreasing incident solar radiation with shade during reproductive growth always reduced soybean seed number (Schou et al., 1978; Egli and Zhen-wen, 1991; Egli, 1993; Jiang and Egli, 1995; Board and Tan, 1995; Egli, 2010; Nico et al., 2015). Shade treatments that lasted for the entire reproductive period (R1 to R8) always had a greater effect than shade during only part of reproductive growth (Table 4.4), however, very short periods of shade (4–9 days) during the peak fruit production period had essentially no effect of final fruit number (Egli, 2010). Interestingly, reducing radiation only during vegetative growth (before R1) decreased plant growth by 34%, but had no effect on seed number for plants growing in 0.38 m rows (Table 4.4). The small

Table 4.4. Variation in radiation levels and seed number in soybean. Adapted from Jiang and Egli (1995).[1]

Treatment	Row spacing (m)	Seed number (no m^{-2})
Control	0.38	1838
	0.76	1848
Shade from[2]		
emergence to R1	0.38	1815NS
	0.76	1650NS
R1 to R4	0.76	1313*
R4 to R6	0.76	1242*
R1 to R8	0.76	810*

*Significant at $\alpha = 0.05$, NS=not significant.
[1]Cultivar Pennyrile, average of 1991 and 1992.
[2]Shade cloth reduced incident radiation by 63%.

effect of pre-flowering shade on seed number of plants growing in 0.76 m rows is probably related to lower radiation interception during flowering and pod set caused by reductions in leaf area, i.e. an indirect effect of the pre-flowering environment. Defoliation during Stage Two also reduced seed number (Board and Tan, 1995).

Soybean seed number is also sensitive to moisture stress during flowering and podset (Shaw and Liang, 1966). Pod number was affected most during the early stages of flowering and podset, while seeds per pod was affected only by stress near the end of the pod set period. Hardman and Brun (1971) reported increases in nodes, pods and seeds per unit area when soybean was grown in a CO_2-enriched environment during flowering and podset. Increasing solar radiation during flowering and podset increased pod and seed number at maturity in a two-year field study (Schou *et al.*, 1978). High-temperature stress near the end of the pod set period reduced seed number (Spears *et al.*, 1997) as did N stress (Brevedan *et al.*, 1978).

Soybean seed number clearly responds to manipulations of the environment that affect canopy photosynthesis during flowering and fruit set. Stress (shade, lack of water and N, high temperature) reduced seed number and improved environments (CO_2 levels above normal, increased radiation) increased seed number.

Seed number in wheat responded to increased radiation levels created by removing border rows at different times during vegetative and early reproductive growth (Fischer and Laing, 1976). The earliest thinning treatment (approximately 60 days before anthesis) increased seed number, but there was little effect when thinning occurred at anthesis (Fig. 4.1). As expected from the sequential nature of sink formation, the number of spikes accounted for most of the increase from early thinning, with a smaller contribution from spikelets per spike, while seeds per spikelet was affected only by treatments closer to anthesis. Thinning after anthesis (i.e. out of the critical period) had no effect on seed number.

Reducing canopy photosynthesis of wheat by reducing solar radiation before anthesis reduced seed number (Fischer and Stockman, 1980; Slafer *et al.*,

1994b), as did N stress (Blacklow and Incoll, 1981; Jeuffroy and Bouchard, 1999; Oscarson, 2000) and high temperatures (Rawson and Bagga, 1979). Krenzer and Moss (1975) increased seed number by increasing CO_2 concentrations to 600 μl l^{-1} from floret initiation to anthesis, with two cultivars in the field. Seed number in wheat, like soybean, is responsive to a variety of changes in its environment.

Maize plants lost their ability to increase ears per plant in response to increases in photosynthesis (created by reducing plant density) at roughly 50% silk emergence (Fig. 4.5).

Seeds per plant responded to increases in solar radiation per plant from two weeks before until two weeks after silk emergence (Schoper et al., 1982). Water stress and shade treatments shortly before and after pollination reduced seed number (Hall et al., 1981; Uhart and Andrade, 1995a) as did N stress (Uhart and Andrade, 1995a). Maize exhibits the same response already established for soybean and wheat; seed number responds to changes in the plant's environment during Stage Two.

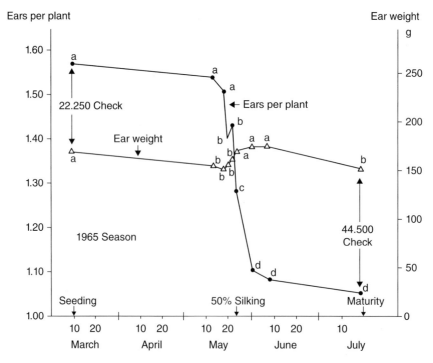

Fig. 4.5. The effect of thinning to increase photosynthesis per plant at various times during the growth and development of maize on ears per plant and weight per ear at maturity. Plots were thinned from 44,500 to 22,500 plants ha^{-1}, while the checks were at the appropriate population for the entire growth cycle. Means followed by different letters are significantly different at $p < 0.05$. From Prine (1971).

Seed number in these three crop species responded to a variety of changes in the plant's environment including temperature, water availability, CO_2 concentration, solar radiation and soil fertility. Similar responses to the environment have been reported for many other grain crops, including chickpea (Lake and Sadras, 2014), sunflower (Lindstrom *et al.*, 2006) and rice (Yang *et al.*, 2009; Kobata *et al.*, 2013). Changes in the environment that would be expected to improve canopy photosynthesis and crop growth (high CO_2 concentrations, more radiation, high levels of N) increased seed number while changes that should reduce canopy photosynthesis and crop growth (low N, water stress, reduced radiation) reduced seed number consistently for soybean, maize and wheat and other crops. There is ample evidence that all of the components of seed number (see Equation 4.7, 4.9, 4.10) respond to these changes in environmental conditions with the component affected determined by the growth stage when the treatment was applied. These responses occurred in crops that are quite diverse including C_3 and C_4 species, legumes and non-legumes, and seeds that are high in starch or high in protein and oil. The consistency of these environmental responses suggests that a common mechanism is controlling seed number in all grain crops.

The common element among all of the environmental factors effects described previously is that they all affect photosynthesis, suggesting that the availability of assimilate from photosynthesis controls the survival of reproductive structures and determines final seed number. Canopy photosynthesis is directly related to radiation levels, photosynthesis of C_3 species increases as ambient CO_2 levels increase, changes in plant population affect solar radiation and photosynthesis per plant; photosynthesis is related to tissue N levels and water stress usually decreases photosynthesis. It is easy to argue that photosynthesis is the physiological process that senses changes in the productivity of the environment and mediates the response of seed number. Invoking photosynthesis as the process responsible for determination of seed number provides a mechanism that allows totally different environmental factors, unrelated in their physical or physiological basis, to have the same effect on seed number and its components. The mechanisms by which these environmental factors affect photosynthesis are different (i.e. stomatal closure, energy supply, substrate availability), but the effect on seed number is the same. Following this scenario, any change in the plant's environment that influences photosynthesis during Stage Two will affect seed number. Environmental interference with pollination or a sink limitation (potential seed number is less than the number that could be supported by the assimilate supply) are exceptions to this rule.

Relating seed number to photosynthesis and the availability of assimilate establishes a link between the productivity of the plant or crop and its reproductive potential (i.e. seed number). Such a relationship makes it possible for the plant to maximize its reproductive output for any level of productivity. In evolutionary terms, the processes during Murata's Stage Two operate to maximize reproductive fitness and the chances of survival of viable offspring; from the viewpoint of modern cropping systems, this relationship maximizes yield potential in any environment. Interestingly, maize, a highly productive crop, fits this scenario only when it is managed properly by increasing plant population until there are an excess number of flowers per unit area (Egli, 2015b).

Since the availability of assimilates determines seed number, simply increasing seed number will not increase the availability of assimilate and would not increase yield (Egli and Bruening, 2003). The only exception is when yield is limited by the number of seeds, i.e. it is sink-limited. Any argument that yield can be increased by simply producing more seeds by, for example, artificially reducing flower and pod abortion runs completely counter to the evolutionary goal of matching reproductive output to the productivity of the environment and our argument that the availability of assimilate determines seed number.

There are some situations, however, where seed number is obviously not related to photosynthesis and the assimilate supply. High temperature (e.g. maize, Herrero and Johnson, 1980; rice, Satake and Yoshida, 1978; cowpea, Warrag and Hall, 1984), or moisture stress (e.g. maize, Bolanos and Edmeades, 1996) can disrupt pollination or fertilization as can boron deficiency in wheat (Rawson, 1996) or low temperature in rice (Murata and Matsushima, 1975). In these situations, seed number will be reduced, but this reduction cannot be attributed to a lack of assimilate. These disruptions of fertilization in stress environments can be very important when they occur, but they represent only isolated exceptions to the general relationship between seed number and photosynthesis. It is also possible that seed number may be limited by the number of flowers, not by assimilate availability, as occurs when maize is grown at populations that are too low (Egli, 2015b). Again, this limitation can be important when it occurs, but it is normally not an issue in modern production systems.

The simple hypothesis that seed number is determined by photosynthesis during the critical period is consistent with a large body of literature, but the basic physiological mechanism governing this relationship remains elusive. Is the supply of sucrose or some other carbohydrate to the developing reproductive sink the key factor or is reproductive survival controlled by a process that is only indirectly linked to photosynthesis? Proposed mechanisms include the concentration of assimilate in the phloem (Wardlaw, 1990), the ratio of hexose sugars to sucrose in the seed (Weber *et al.*, 1998), regulation of the rate of transfer of assimilate to the seed (Liu *et al.*, 2004; Ruan *et al.*, 2012), or assimilate regulation of genes causing abortion (Boyer and McLaughlin, 2007). Hormones have been linked to flower abortion in soybean (Huff and Dybing, 1980), but if they play a regulatory role, it must be related to photosynthesis. The simplest mechanism, and one that is consistent with a large body of experimental observations, relates seed number to the availability of assimilate, so we will investigate models based on this hypothesis in the next section. There are undoubtedly other models that could be investigated, but the best approach is to start with the simplest model, abandoning it only when the facts require it. Understanding the exact mechanism coupling photosynthesis and seed number would be very useful, but we can use the photosynthesis–seed number relationship to better understand the yield production processes, without knowing the basic mechanism(s) involved.

Modelling seed number and assimilate supply relationships

Understanding the relationship between the assimilate supply and seeds per unit area is important, since most of the environmentally induced variation is yield is associated with variation in seed number (Fig. 4.2–4.4). Realistic models of this relationship will no doubt enhance our understanding of plant–environment interactions regulating this important yield component and thus yield. These models are a key component of most crop simulation models (Ritchie and Wei, 2000). A useful mechanistic model of the determination of seed number by grain crops should, first, encompass all of the environmental factors that influence seed number in crop plants, and, second, account for genetic differences in seed number within and among species.

A model that meets these two requirements would account for all factors known to affect seed number. As discussed previously, the determination of seed number starts early in plant development, with the development of the primordia of the structures that bear the seeds, but since the determination of seed number is usually a downward adjustment process from potential to the final number, it is not necessary to model potential seed number. All models of seed number follow this convention. The adjustment in soybean and probably other grain legumes is determined by the number of pods that survive to produce mature seeds. In other species, survival of individual seeds is probably more important. In some species, survival of other reproductive structures (spikelets, tillers) may be more important. Regardless of this dichotomy, we will focus on seed number as we investigate models relating seed number to the assimilate supply. At first glance it may seem that a very complicated model will be required to accommodate both environmental and seed characteristics. As discussed previously, however, photosynthesis provides an integrative mechanism to relate all aspects of the environment to seed number.

Simple correlative models

Early models were based on simple correlative relationships between some measure of crop productivity and seed number. Stapper and Arkin (1980) related seed number in maize to biomass per unit area at maturity, a representation of productivity integrated over the entire life cycle of the crop. Close associations between crop growth rate (an estimate of canopy photosynthesis) during Murata's (1969) Stage Two and seed number have been reported for soybean (Fig. 4.6, Herbert and Litchfield, 1984; Ramseur *et al.*, 1985; Egli, 1993: Jiang and Egli, 1995), maize (Hawkins and Cooper, 1981; Cirilo and Andrade, 1994; Uhart and Andrade, 1995a), rice (Cock and Yoshida, 1973), sorghum (Gerik *et al.*, 2004), groundnut (Phakamas *et al.*, 2008). Similar relationships were reported by Charles-Edwards *et al.* (1986, p. 125) using data from Pandy *et al.* (1984a,b) for four grain legumes (cowpea, soybean, mungbean (*Vignia radiata* L.) and groundnut) where variation in crop growth rate was created by differences in water supply. These relationships no doubt exist for all grain crops and they support a close association between canopy photosynthesis during the critical period for seed number and

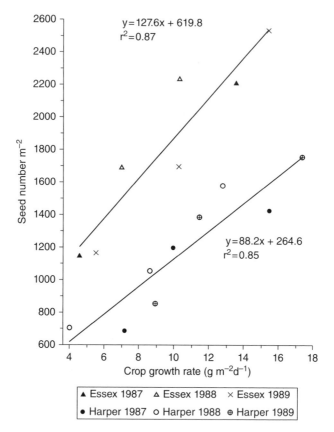

Fig. 4.6. The relationship between seeds m⁻² and crop growth rate for soybean cultivars that differ in individual SGR, 1987–1989. Mean individual SGRs were 6.3 mg seed⁻¹ day⁻¹ for Harper and 4.5 mg seed⁻¹ day⁻¹ for Essex. Adapted from Egli and Zen-wen, 1991.

the number of seeds produced by the plant community. This substantial body of data strongly supports the basic premise expressed earlier that seed number is a function of canopy photosynthesis in grain crops.

Several authors successfully related seed number to intercepted photosynthically active radiation (PAR) during Stage Two (Hashemi-Dezfouli and Herbert, 1992; Kiniry and Knievel, 1995). Although intercepted PAR is related to canopy photosynthesis, it does not capture the variation in assimilate availability that could occur without any variation in intercepted PAR (Otegui and Andrade, 2000).

Fischer (1985) expanded the correlative relationship between seed number and intercepted radiation by including temperature-induced variation in the length of the critical period. The photothermal quotient (Equation 4.11) (PQ, MJ m⁻² day⁻¹ °C⁻¹), was defined as the mean intercepted radiation during the critical period (MIR) divided by the difference between the mean temperature during the critical period (MT) and the base temperature (BT, i.e. the temperature where the rate of development is zero).

$$PQ = MIR(MT - BT)^{-1} \tag{4.11}$$

The exact formulation varies (Oritz-Monasterio *et al.*, 1994), but, in all versions, increasing temperature reduces PQ for any level of mean intercepted radiation, i.e. the length of the period is important. A number of researchers reported that relationships between PQ and seed number across locations and years were better than just using intercepted solar radiation (Fischer, 1985; Ortiz-Monsatario *et al.*, 1994 – wheat; Cantagallo *et al.*, 1997 – sunflower; Poggio *et al.*, 2005 – pea; Islam and Morison, 1992 – rice).

Correlative models relate seed number to some average measure of crop productivity, so they do not capture the dynamics of the system, when seeds are produced, how long Stage Two is and potential changes in assimilate supply during the critical period. The importance of this variation is not well understood, but modelling efforts suggest that it could affect final seed number (Egli, 2015a) and could contribute to the failure of some models to produce accurate predictions of seed number in a wide range of environments.

Correlative models do not include the characteristics of the individual seed, which also influence seed number (Fig. 4.6) (Charles-Edwards *et al.*, 1986, p 125). Seed characteristics account for the large differences in seed number among species and can be important when there are genetic differences in SGR and seed size among cultivars within a species. The failure to include sink characteristics is an important limitation of correlative models.

Complex models of seed set

The models developed by Sheldrake (1979) and Duncan (W.G. Duncan, unpublished manuscript, 1982) moved beyond the correlative approach to consider a more mechanistic relationship between the availability of assimilate and the potential survival of individual fruits and seeds. The availability of assimilate in Sheldrake's (1979) hydrodynamical model was represented by the size of the water reservoir (the source) and the flow of water from the reservoir to the reproductive 'sink' shown in Fig. 4.7. Water in the reservoir could flow into the reproductive 'sink' only if the water depth in the reservoir was greater than the threshold level required to initiate flow to the sink. The depth of water in the reservoir depends on the rate of addition (canopy photosynthesis) and the flux to the sinks. Sheldrake (1979) introduced the concept that a minimum assimilate supply (threshold level, Fig. 4.7) must be available before sink development can proceed.

W.G. Duncan proposed a similar threshold concept for seed set in soybean in a paper presented at a meeting of soybean researchers in St Louis, Missouri in 1982 (Fig. 4.8). Duncan's model applied the threshold or minimum flux concept only during the initial stages of seed development. Valve B is closed during the initial critical period and the seed receives no assimilate (i.e. it aborts) if the assimilate supply is not adequate to provide flow over the threshold. After the initial critical period, valve B (Fig. 4.8) opened and assimilate flowed directly to the

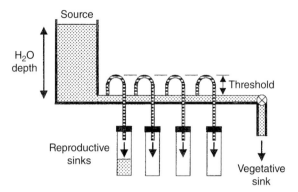

Fig. 4.7. A hydrodynamical model of the relationship between photosynthesis and reproductive sink size (fruit or seed number). The water in the reservoir (size of the reservoir and the depth of the water) represents the supply of assimilate. Water will not flow into the sinks unless the depth in the reservoir is greater than the threshold. The rate of flow is governed by the water depth and the size of the tube connecting the sink to the source. Adapted from Sheldrake (1979) and Wardlaw (1990).

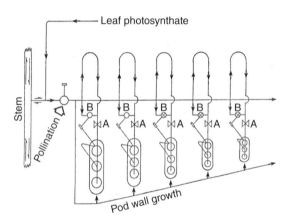

Fig. 4.8. Model of seed set in soybean proposed by W.G. Duncan (unpublished manuscript, 1982). The model represents a single node of a soybean plant. Assimilate flows to vegetative growth until pollination opens the valve and initiates flow to the developing fruits. The hydraulic gradient must exceed the threshold level (represented by the loops above the main axis of the diagram) to initiate flow to the seed. Valve A regulates the rate of flow to the fruit and valve B opens and bypasses the threshold loop when the seed is past the critical period and will no longer abort. The rate of flow through valve A approximates the combined SGR of all seeds in the fruit.

seed, bypassing the threshold loop, so that the seed always receives some of the available assimilate. After the initial critical period, the seed will not abort. Pod growth did not involve the threshold or Valve B. Duncan recognized that it is necessary to divide fruit or seed development into two phases, an initial phase when growth could not continue without a threshold level of assimilate and a second phase where growth continued at all levels of assimilate availability. Subsequent research confirmed that legume fruits or seeds are impervious to abortion after the fruit reached its maximum length (Heitholt *et al.*, 1986; Duthion and Pigeaire, 1991; Egli and Bruening, 2006b), which is roughly when the seeds enter the linear phase of growth.

Both of these models implicitly include the characteristics of the seed sink as determined by the magnitude of the threshold, the size of the tubing or the state of valve A connecting the sink to the source (Fig. 4.7 and 4.8). For a given assimilate supply, the number of fruits or seeds set would be inversely related to the magnitude of the threshold or the utilization of assimilate by the individual seed or fruit. Both models also imply that the timing of seed or fruit development (position of the fruit relative to the source in Fig. 4.7 and 4.8) plays a role in determining which seeds survive. Again, subsequent research has shown that the first seeds to develop have a much higher probability of surviving than later developing fruits (Huff and Dybing, 1980; Heitholt *et al.*, 1986; Egli and Bruening, 2006b).

Charles-Edwards developed a simple, but elegant, mathematical model to describe the relationship between the number of vegetative meristems or reproductive plant parts and photosynthesis. This model was first described briefly in Charles-Edwards (1982, p. 103), and then in more detail in a series of papers entitled 'On the ordered development of plants' (Charles-Edwards, 1984a, b; Charles-Edwards and Beech, 1984). Charles-Edwards' model also related the number of sinks to the availability of assimilates and it included the minimum flux (rate of supply) concept. A minimum supply of assimilate was required for continued development of the sink; if the flux dropped below the minimum, development ceased, and the sink aborted. The magnitude of the minimum flux was a characteristic of the sink (seed), involving the sink in the determination of seed number. The total flux of assimilate to the developing sink was determined by the proportion of canopy or individual plant photosynthesis that was partitioned to the developing sink. Since Charles-Edwards (1982) formalized his model in an equation, we will focus on his model for the rest of this discussion.

Charles-Edwards' concept, applied to reproductive plant parts, is described by Equation 4.12 (Charles-Edwards *et al.*, 1986, pp. 124–127) where:

$$N_G = \eta_G \nabla_F / a_G \qquad\qquad (4.12)$$

N_G = number of floret primordia or seeds per unit area
η_G = proportion of current assimilate partitioned to reproductive plant parts
∇_F = canopy daily net photosynthetic integral per unit area
a_G = minimum assimilate flux requirement of an individual floret primordia or seed

In this iteration of the model, N_G was described as floret primordia or seeds (Charles-Edwards *et al.*, 1986, pp. 124–127); however, earlier descriptions of the model (Charles-Edwards, 1982) make it clear that N_G could describe all reproductive and vegetative sinks, including branches and roots (Charles-Edwards, 1984a). Equation 4.12 includes all of the components found in the Sheldrake and Duncan models, i.e. canopy photosynthesis (∇_F), partitioning (η_G), and the characteristics of the sink (minimum flux of assimilate (a_G) needed to maintain development of an individual sink). All of these models make it clear that the number of reproductive sinks is not simply a function of assimilate available for reproductive growth (i.e. photosynthesis (∇_F) and partitioning (η_G)), but it is also affected by the characteristics of the sink. The inclusion of sink characteristics (a_G) in the model suggests that they are not constant and perhaps a_G is determined by the type of sink, cultivar, species, or even by the environment. This term provides another avenue, a very important avenue, as we shall see, for the expression of factors causing variation in seed number that are not related to the productivity of the crop. Charles-Edwards' model also included the concept that the reproductive sink was sensitive to the assimilate supply only during initiation and establishment and thereafter represented a 'passive' sink for new or remobilized assimilate. This concept, however, was not included in Equation 4.12.

These complex models include many of the factors known to influence seed number' consequently, they represent a significant improvement over the simple correlative models. They do not, however, explicitly include time, but it is clear that the determination of seed number occurs throughout Murata's Stage Two (Murata, 1969). Without a time component, the Charles-Edwards model represents seed number at maturity as determined by average values for the variables in the model. Canopy photosynthesis and possibly partitioning and other important variables could exhibit day-to-day variation during Stage Two, which could influence seed number.

Flower production and fruit and seed development is a highly dynamic process in most crop plants and the availability of flowers and fruits or seeds and their stage of development at any time can be an important determinant of seed number. The temporal pattern of flower production varies widely among species, with soybean, for example, producing flowers for 30–40 days (Egli and Bruening, 2006a), while flowering in other species, such as maize and cereals, is much more compact (Evans *et al.*, 1972; Tollenaar and Daynard, 1978). This temporal variation results in each fruit developing in a potentially different environment that may influence the availability of assimilate to that fruit. Temporal variation also influences competition among developing fruits for scarce assimilate. The critical period in the development of each flower/fruit would also interact with the temporal variation in assimilate availability. Synchronous fruit development often increases pod or seed set (Freier *et al.*, 1984; Carcovca *et al.*, 2000; Egli and Bruening, 2002) and it is often the late-developing fruits that abort (Heitholt *et al.*, 1986). These observations demonstrate that the dynamic nature of flowering and fruit or seed development is important and its inclusion in models of seed set could improve their sensitivity to short-term fluctuations in environmental conditions; conditions that are not captured by models relating an average estimate of assimilate availability during the critical period to seed number at maturity.

A dynamic model of seed set in soybean, SOYPODP (Egli, 2015a), related the survival of individual fruits to the availability of assimilate during the critical period of development of each fruit. The fruit dry matter accumulation rate and the temporal pattern of flower production at each node were taken from measurements on field-grown plants. A fruit aborted and was no longer a sink for assimilate when there was not enough assimilate available to meet its growth requirement during its critical period. If there was not enough assimilate to supply all fruits, assimilate was distributed to the fruits in order of their age (oldest fruits had top priority). The model accurately reproduced the distribution of fruits on the main stem and the typical response of fruits per plant to variation in assimilate availability noted in field experiments.

Manipulation of the parameters controlling the dynamics of flowering and fruit development in SOYPODP resulted in variation in fruits plant[-1] at a constant level of assimilate availability. Shortening the flowering period at each node increased fruits per plant, primarily as a result of decreasing competition among fruits for limited supplies of assimilate (Fig. 4.9). This response from a relatively simple model suggest that capturing the dynamics of flower production and fruit development may be necessary to accurately model fruit and seed numbers in a dynamic environment. The implications of increasing fruit set per unit assimilate on yield will be discussed in Chapter 5.

Correlative models often produce excellent relationships between seed number and plant growth, but they are not mechanistic or sensitive to plant characteristics, so they may not capture important nuances of the seed number

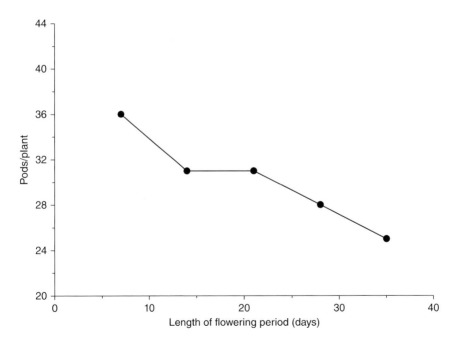

Fig. 4.9. Simulating the effect of varying the length of the flowering period on pods per plant with SOYPODP. The simulated plant had 21 nodes on the main stem and 68 flowers per plant. Adapted from Egli (2015a).

determination process. The Charles-Edwards model is more mechanistic and includes seed characteristics, but it does not include time or the temporal dynamics of assimilate and flower production, and fruit and seed growth.

Implicit in all these models is the assumption that potential seed number is greater than the actual seed number; none of them include the process of flower production. Potential seed number (i.e. the number of flowers) must be large enough to accommodate the number of seed determined by assimilate availability and sink characteristics. The relatively high level of reproductive failure in many crops suggests that this assumption is usually satisfied, with maize representing an exception to the general rule. Maize grown at a low plant density may not produce enough flowers to accommodate the available assimilate (Andrade *et al.*, 1993; Uhart and Andrade, 1995a; Vega *et al.*, 2001). Maize did not come by this limitation naturally; it was created by plant breeders favouring single-eared, non-tillering plants, resulting in a reduction in its capacity to increase potential seed number in favourable environments.

Many of the complex models include a provision for partitioning only part of the assimilate from photosynthesis to reproductive growth. Intuitively this is an important aspect of seed number determination because some assimilate must be used for vegetative plant growth, including respiration, nutrient acquisition etc, reducing the amount available for reproductive growth. Partitioning is also a process that defies mechanistic modelling; it is difficult to accurately measure and the processes controlling it remain a mystery. Much remains to be learned about this process before it can be realistically included in models of seed number determination.

The supply of assimilate that determines seed number in these complex models is based solely on current photosynthesis and an arbitrary partitioning factor, no provision is made for utilization of reserve assimilates (starch, fructans, sucrose). These storage carbohydrates accumulate in most crops, and they may even accumulate during Stage Two, when seed number is determined. Including storage reserves in the assimilate supply may make it more difficult to balance the size of the yield container (primarily seed number) and the ability of the crop to fill the container. Storage reserves accumulate over time, so the size of the reserve pool is not necessarily related to daily canopy photosynthesis, which is the primary source of assimilate during seed filling. A crop with a relatively slow rate of canopy photosynthesis could have a relatively large pool of storage carbohydrates if it had a long vegetative growth phase (Egli, 1993, 1997); involving storage reserves in the determination of seed number would probably result in a yield container that is too large, relative to the capacity of current photosynthesis during Stage Three to fill it. Separating seed number from the crop growth rate or the rate of canopy photosynthesis would not match sink size with the capacity of the plant canopy to fill the sink. There is little evidence that seed number is influenced by storage reserves (Egli, 1993; Schussler and Westgate, 1994; Bruening and Egli, 2003). It seems that the assumption in all models that seed number is a function of current photosynthesis is still valid. Ignoring storage reserves is theoretically the best approach and there is little direct experimental evidence available to contradict it.

The models we discussed have no provision for the activity of the sink to af-fect the activity of the source, i.e. feedback control of photosynthesis. Allowing the size of the sink to regulate photosynthesis completely invalidates models that base seed number on current photosynthesis. Including feedback control would require a totally different mechanism for determining seed number. Feedback control has been reported (see Evans (1993, pp. 173–178) for a thorough summary), but there is still no clear cut answer in the literature on the importance of feedback con-trol on photosynthesis in field communities. Artificially reducing reproductive sink size often reduces photosynthesis (Lawn and Brun, 1974; Mondal *et al.*, 1978; Wittenbach, 1983), but increasing sink size relative to the source is much more relevant to the issue here. Experiments where seed number was increased artifi-cially had no effect on yield (Hardman and Brun, 1971; Ackerson *et al.*, 1984) or photosynthesis during seed filling (Egli and Bruening, 2003). Theoretical consider-ations and the consistent relationship between estimates of canopy photosynthesis and seed number suggest that ignoring possible feedback from sink size to source activity is the appropriate approach.

A direct relationship between some estimate of canopy photosynthesis and fruit and seed number is the basic tenet of all models and it provides a reasonably good estimate of the number of mature fruits or seeds for most crop species. Some form of this relationship is used in most crop simulation models. Including the sink characteristics in a model improves the model, making it possible to account for species or cultivar differences in seed size that are related to seed growth rate. Many models do not include temporal variation in flower production and fruit or seed development, the location of the flower in the fruiting structure or on the plant, the length of the flowering period or the effects of daily variation in the as-similate supply. There is evidence from some models suggesting that these aspects of seed number determination are important, but there is also evidence suggesting that fruit or seed survival is not affected by the assimilate supply on a particular day (Egli, 2010, 2015a). It is not yet clear how much improvement would result from including any of these components in a crop simulation model.

Refining and improving models of seed number determination may not be possible until the exact mechanism that causes an individual flower, fruit or seed to survive or abort is known and understood (Egli, 2015a). The mechanism cannot be modelled until we can describe it, and until that happens our models are, in the ultimate sense, simply correlative models, regardless of their complexity. The models, however, describe the main features of the system and, in spite of their weaknesses, help us understand the processes whereby seed number of grain crops is determined.

Determination of Seed Size

Seed size (weight per seed) is the final component of yield and it is determined during Murata's third stage as 'the production, accumulation and translocation of yield contents' fills the yield container (Murata, 1969). Seed size is under genetic control and it is also influenced by the supply of assimilate from the plant during the seed-filling period. Potential seed size, controlled by the characteristics of the fruit and/or the seed, provides the ultimate limit of seed size. Because the determination of seed size is last in the sequential process of yield production, some of the variation in size reflects adjustment to component levels fixed earlier in the sequence (i.e. seed number) and changes in the environment (Evans, 1993, pp. 260–264). Environmental-based adjustments in seed number during Stage Two create a balance between source and sink, which tends to minimize variation in seed size. This balance can be disturbed by changes in environmental conditions from Stage Two to Stage Three, resulting in direct effects on seed size. Persistence in weather conditions in the field during reproductive growth probably reduces variation in seed size, but it is not reduced to zero (see, for example, Fig. 4.2 and 4.3).

Seed size has usually received more attention from agronomists than other yield components, probably because it is so readily observable. From ancient times, farmers were probably well aware of the size of the seeds they harvested, saved and planted. In fact, seed size often increased as crops were domesticated (Evans, 1993, pp. 96–98), when early farmers selected large seeds to save for next year's planting. Large genetic differences exist within and among species (the range in size among species in Table 3.1 is from 7 to more than 2000 mg seed^{-1}) and there is also significant variation among environments.

Potential seed size

The term 'potential seed size' describes the concept that each seed has a maximum size that cannot be exceeded, regardless of assimilate availability. Final seed size may be equal to or less than the potential seed size, but it cannot, by definition, exceed potential size. Potential seed size is a simple concept that intuitively reflects reality; after all, a wheat seed will never be as large as a soybean seed, but deciding whether a seed has reached its potential size is more difficult.

Potential size is set by the size of the fruit or other seed structures (ovary, carpel, glumes, pericarp) that complete their development before there is any significant accumulation of dry weight in the seed. These structures can limit the capacity of the seed to increase in volume, which is closely associated with final seed size (see the discussion of regulation of seed-fill duration in Chapter 3). Relationships between the size of seed structures and seed size have been reported for many species, including legumes (Corner, 1951; Duncan et al., 1978; Frank and Fehr, 1981; Fraser et al., 1982a) and cereals (Murata and Matsushima, 1975; Jones et al., 1979; Scott et al., 1983; Calderini et al., 1999). Physically reducing the capacity of the seed to expand during development reduced seed size (Boshankian, 1918; Murata and Matsushima, 1975; Grafius, 1978; Millet and Pinthus, 1984; Millet, 1986; Egli et al., 1987a; Miceli et al., 1995), supporting the contention that fruit and seed structures can influence seed size.

Environmental conditions and assimilate availability during the early stages of fruit and seed development can influence the size of these structures (Calderini and Reynolds, 2000), suggesting that potential seed size of a cultivar could vary among environments or even among seeds on the same plant, depending on their location or when they developed. Artificially reducing sink size after the early stages of fruit and seed development did not eliminate variation in size of seeds developing at different times or locations on a soybean plant (Egli *et al.*, 1987d), suggesting that there was variation in potential size of those seeds. The number of cells in the embryo or endosperm is often related to final seed size, but the role of cell number in determining potential seed size (i.e. is there a maximum cell volume that cannot be exceeded?) or whether other seed structures are more important is not yet clear.

Involving potential seed size in the analysis of the production of yield is more difficult than expected for such a simple concept. Obviously, potential seed size will limit yield if there is assimilate available during seed filling to support seed sizes that are larger than the potential size. Evaluating such a limitation requires a measure of potential seed size, which is difficult because it requires demonstrating that the seed cannot get any larger. An increase in seed size following sink-reduction treatments (enhancing the source relative to the sink) demonstrates that seed size on the untreated control plant was less than potential, but size of those seeds does not necessarily provide an estimate of potential size. Potential seed size is a useful concept, in spite of our inability to precisely estimate its magnitude. The regulation of potential seed size and its involvement in the yield-production process no doubt varies among species as determined by the structure of the seed, but this variation does not limit the usefulness of the concept.

Components of seed size – seed growth rate and seed-fill duration

Dividing seed size into its components, SGR and SFD (Equation 4.13), provides a useful framework to evaluate the processes regulating seed size.

$$\text{Seed Size} = (\text{SGR})(\text{SFD}) \tag{4.13}$$

Final seed size is simply a function of how fast the seed grows and how long this growth continues. The mechanisms regulating seed size are thus the mechanisms regulating SGR and SFD as described in Chapter 3 and both of these components contribute to the observed variation in seed size.

Seed growth rate and seed size

Much of the variation in seed size is related to seed growth rate. Genetic differences in seed size are usually determined by SGR; large seeds usually have higher SGRs than small seeds leading to a close association ($r = 0.81$) between size and SGR (Fig. 3.2, Egli, 1981). Egli *et al.* (1981) reported a linear relationship between seed size and SGR for seven soybean cultivars (Fig. 4.10). Others have reported similar results for soybean (Egli *et al.*, 1978a; Guldan and Brun, 1985; Swank *et al.*, 1987), maize (Carter and Poneleit, 1973; Reddy and Daynard, 1983; Jones *et al.*, 1996), wheat (Jenner and Rathjen, 1978; Chojecki *et al.*, 1986), cowpea (Lush

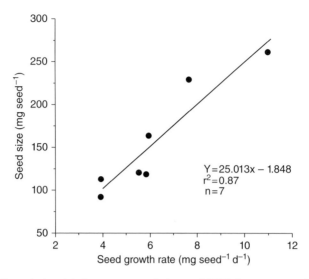

Fig. 4.10. The relationship between seed size and SGR for seven soybean cultivars. Adapted from Egli *et al.* (1981).

and Evans, 1981) and common bean (Sexton *et al.*, 1994). These differences in rate are under genetic control and they are related to the number of cells in the endosperm or cotyledons, as described in Chapter 3. By far the great majority of genetic or species differences in seed size result from differences in SGR and these differences in SGR are regulated by the characteristics of the seed, not by the supply of assimilate to the seed.

Variation in the supply of assimilate to the seed during the seed-filling period also affects SGR and seed size. Modification of the environment with shade to reduce soybean canopy photosynthesis during seed filling reduced SGR (Table 3.3) and seed size (Table 4.3). Reducing seed number to increase the supply of assimilate to the remaining seeds, increased SGR and seed size in soybean (Fig. 3.4) (Egli *et al.*, 1985a), maize (Kiniry *et al.*, 1990; Borras *et al.*, 2003), sorghum (Kiniry, 1988; Gambin and Borras, 2007) and wheat (Table 3.3; Slafer and Savin, 1994). The saturation response of SGR to sucrose concentration discussed in Chapter 3 (Fig. 3.3) suggests that seeds may be more likely to respond to a reduction in assimilate supply than to an increase making seed size more sensitive to environmental stress than to an improved environment during seed filling. The failure of SGR to increase in some experiments when sink size was reduced (Egli *et al.*, 1985a; Munier-Jolian *et al.*, 1998) is consistent with this suggestion.

Variation in seed size by location on the plant or in the reproductive structures can also be a result of variation in SGR. Often the late-developing seeds are smaller and have lower SGRs than those from earlier flowers, as shown for wheat (Miralles and Slafer, 1995), rice (Kato, 1986) maize (Tollenaar and Daynard, 1978; Frey, 1981), sorghum (Gambin and Borras, 2005) and sunflower (Lindstrom *et al.*, 2006). Small seeds from late-developing flowers in soybean, however, did not consistently exhibit lower seed growth rates (Egli *et al.*, 1978a; Gbikpi and Crookston, 1981), so not variation in seed size associated with location and time of development effects is all due to changes in SGR.

Seed-fill duration and seed size

Variation in SFD also makes a contribution to genetic differences in seed size, as shown by a significant correlation ($r = 0.50^{**}$) between size and SFD across species and cultivars (Fig. 3.2, Egli, 1981). One can easily find substantial differences in seed size among species in Fig. 3.2 at a constant SGR because there is a wide range in SFD (7–57 days). The variation in SFD among commonly grown cultivars of major crops is much less, so the correlation of seed size and SFD within a species would probably be much smaller. The maximum SFD among genotypes within a species was often only 20% longer than the minimum duration (Egli, 2004), but Swank *et al.* (1987) was able to select soybean plant introductions with a large range in seed size (roughly 100–325 mg seed^{-1}) that was primarily due to variation in SFD (Fig. 4.11). Tollenaar and Bruulsema (1988) worked with two maize hybrids with different seed sizes (272 and 234 mg seed^{-1}) that were associated with differences in SFD.

Environmental effects on seed size can also be caused by variations in SFD. For example, water stress reduced seed size by shortening the seed-filling period in soybean (de Souza *et al.*, 1997), barley (Aspinall, 1965), wheat (Ahmadi and Baker, 2001), pearl millet (Bieler *et al.*, 1993), chickpea (Davies *et al.*, 1999) and

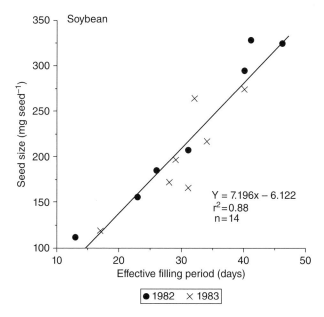

Fig. 4.11. The dependence of seed size on SFD (effective filling period) for seven soybean genotypes grown in the field for two years. The genotypes were selected for variation in seed size and a constant SGR. Adapted from Swank *et al.* (1987).

maize (Jurgens *et al.*, 1978). Nitrogen stress during seed filling in soybean (Egli *et al.*, 1985a; Hayati *et al.*, 1995) also reduced seed size by shortening the seed-filling period. Increasing temperature may shorten the seed-filling period, but the effect on seed size depends on the effect of temperature on SGR (Chowdhury and Wardlaw, 1978; Tollenaar and Bruulsema, 1988). Excluding temperature effects, it is hard to find examples of environmental conditions that increased seed size by lengthening the seed-filling period.

As discussed in Chapter 3, location-specific variation in seed size associated with the timing of seed or fruit development may be related to SFD. Seeds developing late in the flowering period may be smaller and have a shorter seed-filling period than early developing seeds because they both reach physiological maturity at nearly the same time (wheat, Rawson and Evans, 1970; maize, Frey, 1981). The contribution of variation in SFD to differences in seed size varies among crop species; for example, soybean seed size, and presumably SFD, was not closely associated with time of flowering or initial development of individual fruits (Egli *et al.*, 1987d; Egli, 2012). The intra-plant variation in SFD is not well documented for many grain crops, but it is clear that it makes some contribution to the variation in seed size.

Regulation of seed size involves interactions among the supply of assimilate, SGR, SFD and potential seed size. We must not forget that the seed cannot grow without a continuous supply of raw materials from the mother plant, so the ability of the plant to supply these raw materials is always the ultimate control. The assimilate supply has a rate component (supply per day) and a duration component (how long is the supply maintained?). Variation in either of these aspects of supply can, within the constraints of SGR, SFD and potential seed size, influence seed size.

The simplest scenario is a reduction in the assimilate supply (rate or duration) during seed filling, which will almost always reduce seed size of all crops, either by reducing the SGR or shortening the seed-filling period. Potential seed size is not limiting because seed size is reduced, which probably contributes to the consistency of the response. The only exception to the typical response could occur in a crop that was seriously sink-limited during seed filling (i.e. seed size equals potential seed size) and was still sink-limited after the assimilate supply was reduced, so seed size would not change. This response would probably be rare in well managed crop production systems.

Increases in seed size are, however, more complicated and harder to understand. Seed size can respond to an increase in assimilate supply only if SGR can increase and/or seed fill can be extended, but both responses can be limited by potential seed size. Seed growth rate will not respond to the increase in assimilate if it is saturated with assimilate, so any effect on seed size would have to come from a longer seed-filling period and a delay in senescence. Increasing SGR or SFD will increase seed size until it is limited by the potential seed size. A longer seed-filling period is only possible when assimilate supply is maintained and the combination of potential size and SGR are such that continued growth of the seed is possible. A higher SGR could also result in a shorter seed-filling period and no change in size when potential size is limiting (Kato, 1986). On the other hand, if SGR was low relative to maximum seed size, it would take longer for the seed to reach that size, resulting in a longer seed-filling period, as shown for soybean by Swank *et al.* (1987). If potential seed size limits the increase in size, seed maturation could occur before senescence is completed, resulting in plants with mature seeds and green leaves. These potential interactions among SGR, potential seed size and leaf senescence make it difficult to predict the effects of increasing the assimilate supply during the seed-filling period on seed size.

The relationships underlying this interplay of SGR, SFD and potential size are generally consistent across grain crop species, but there are species differences in the details, possibly related to seed structure. One important difference may be the timing of the occurrence of maximum seed volume and potential seed size, which occurs earlier in seed development in maize, wheat and probably other cereals than in soybean and probably other legumes (Borras *et al.*, 2004). Whether occurring later in the seed-filling period provides greater opportunities to respond to increases in assimilate supply by increasing maximum volume and potential seed size is not known. In fact, potential seed size is so poorly understood, it is impossible to determine how often it is involved in limiting the capacity of seed size to respond to increases in the supply of assimilate (rate or duration).

Seed size is ultimately determined by the interaction of the characteristics of the fruit or seed and the supply of assimilate during seed growth and development. These interactions may help explain the variety of responses to source–sink alteration treatments, both within and among crop species. Evaluating seed size from the viewpoint of potential seed size and dry matter accumulation by the seed leads to a better understanding of the processes involved and helps clarify the many interactions between fruit and seed characteristics and the assimilate supply that combine to determine final seed size.

Summary

Dividing yield into its components, seed number and seed size, and investigating the mechanisms that regulate the levels of each component is the key to a better understanding of the yield production process. Plant communities do not produce yield, they produce flowers and seeds, and then the seeds grow to their mature size. Yield (weight of seeds per unit area) is a concept developed by man to judge the productivity of his crops, so it is only tangentially related to the crop processes that produce it. To understand how yield is produced, we have to consider the processes involved in the determination of the yield components – the production of the yield container and filling of that container. The sometimes complex interrelationships of yield components and the problems of yield component compensation do not justify abandoning yield components; in fact, we have to study these components to understand yield. The old idea that there were yield genes was only correct in the abstract; genes control the production of the components and thus yield, but not yield directly.

C.M. Donald made it clear in 1962 that if yield is controlled by photosynthesis and if photosynthesis is constant, change in one component must cause a compensatory change in another component (Donald, 1962). Donald made it clear that increasing seed number or seed size without increasing the availability of photosynthate (either rate or duration) will not increase yield of a crop, a fundamental relationship that is too frequently ignored by crop physiologists and plant breeders. If we understand the regulation of yield components, we will have a better understanding of the production of yield, and yield component compensation will no longer cause consternation and confusion.

Considering yield components also emphasizes the temporal sequence of yield accumulation. Recognizing that the number of seeds produced by the plant community (the yield container) is determined before seed filling (filling the yield container) is critical to understanding the effect of the environment on yield and how to manage grain crops for maximum yield.

By combining yield components and the characteristics of growth and development of the individual seed, we have developed an understanding of the mechanisms that regulate seed number and seed size. In the next chapter, we will use these mechanisms to investigate the relationship between the plant, the environment and yield.

The Seed, Crop Management and Yield

5

Growth and development of a grain crop involves a multitude of physiological and physical processes operating in a highly coordinated fashion against a background of a constantly changing environment. Yield is the product of these processes integrated over 70–100 days or more from planting to maturity. Evaluating this system from the viewpoint of basic metabolic processes is usually disheartening and rarely leads to a clear understanding of the processes regulating yield in the farmer's field. It is often difficult to use knowledge at the level of metabolic process, organelle or cellular to understand the production of yield. Much of what we know about the critical processes controlling yield of grain crops comes from investigations at the whole plant–plant community level.

This empirical approach by agronomists and crop physiologists has advanced practical agriculture; but, unfortunately, it has not always lead to a deeper understanding of the yield production process. Modifying the crop's environment and observing the response is a common approach; less common is asking why the response occurred, the key question that must be answered to increase the portability of the findings and to develop a useful 'model' of the yield production process.

Models of the yield production process range from simple models or equations with only a few terms to complex computer simulation models with thousands of lines of code (e.g. Boote *et al.*, 1998). A simple model often used by crop physiologists describes yield as a function of intercepted solar radiation, radiation use efficiency (dry matter produced per unit intercepted solar radiation) and harvest index (yield/(vegetative mass + yield)). While this model is theoretically correct, it combines almost all important yield production processes into just two terms: radiation use efficiency and harvest index, thereby limiting its explanatory potential. More detailed crop simulation models often deal with many of the processes involved in producing yield and they may have excellent predictive capabilities, but the underlying processes may not be apparent to the casual user.

I think we can develop a useful simple conceptual 'model' of the yield production process at the whole plant–plant community level by focusing on the seed, the plant part that is harvested as yield in grain crops. This 'model' will not be a crop simulation model that 'grows' the crop and estimates yield; rather, it will be a simple description of the processes that are involved in the production of yield. This conceptual model will provide a useful framework to help us understand the yield production process, a framework that deals more explicitly with important yield production processes than the radiation use efficiency–harvest index approach. This framework will inform and guide our research towards a deeper understanding of the yield production process and may help us devise strategies to increase yield.

Agricultural crops that are grown for their seeds are predominantly grasses and legumes, but they are not as homogenous as the groupings imply. Instead, they exhibit variation in photosynthetic pathways, seed composition, morphology, growth habit and their response to environmental conditions. In spite of this variation, I think that one conceptual model can be used to describe all grain crops with variation in the finer details accounting for species differences.

The plant processes responsible for the production of yield by a grain crop can be separated into two categories: those responsible for the production of dry matter from CO_2, water and mineral nutrients, and those responsible for the growth of the seed. This division will be familiar to crop physiologists as the source and the sink. The assimilatory processes are, of course, of primary importance because they are responsible for the capture of solar energy and the fixation of carbon into organic compounds for growth. Without these assimilatory processes, there can be no growth and yield production. The raw materials supplied by the assimilatory processes of the mother plant are utilized by the seed to produce the materials that make up yield and give the seed its value.

The capacity of a crop community to fix carbon is relatively well understood. The environment must supply solar radiation, CO_2, adequate water, a temperature that is suitable for plant growth, and adequate supplies of the necessary mineral nutrients. The proportion of the incident solar radiation intercepted (a function of LAI and canopy characteristics) and the basic photosynthesis processes (C_3 vs C_4) also play important roles in community productivity. An environment that is free from predators, disease, competition from weeds and toxic materials facilitates functioning of the assimilatory processes.

The functioning of the assimilatory processes at the community level is well documented in many crop physiology texts (e.g. Hay and Porter, 2006). Our understanding of the operational efficiency of the system, the respiratory costs of growth, how this cost is influenced by the environment, plant defensive mechanisms, or root exudates is not as well understood. The effects of the environment on these efficiency factors may limit the ability of many models to adjust to changes in locations and environments. Generally, these processes are well enough understood at the community level to make it possible to predict community dry matter accumulation with some accuracy. If economic yield is the entire plant or the above ground portion of the plant, as it is with forages, tobacco (*Nicotiana tabacum* L.) or many bioenergy production systems, we need only understand these assimilatory processes to understand yield. Predicting seed yield is, unfortunately, more difficult than predicting total biomass.

Economic yield of a grain crop is the seed, which represents only a fraction of the total biomass. Production of yield involves, in addition to the primary assimilatory processes, those processes involved in the determination of seed number and potential seed size (combine to determine the size of the yield container, i.e. sink size) and the accumulation of storage compounds in the seed (filling the container). Including the seed as part of the yield production process is crucial to a complete understanding of how yield is produced. Approaching yield from the viewpoint of the seed, instead of focusing exclusively on the assimilatory processes, makes it possible to develop a deeper understanding of the production of economic yield, an understanding that is much richer than simply defining seed yield as an empirical fraction of total biomass.

The seed plays an important role in the production of yield primarily because the characteristics of the seed control, in part, the accumulation of dry matter by the seed as discussed in Chapter 3. The seed is not simply a receptacle for assimilate produced by the leaves, it is a metabolic factory producing complex storage materials from simple raw materials provided by the mother plant. The ultimate size of the seed (the quantity of storage materials produced), is determined by seed growth rate (SGR) and the seed-fill duration (SFD) within the restrictions set by potential seed size; all of which are at least partially determined by the characteristics of the seed.

Understanding seed growth and development provide a mechanistic description of how seed number and seed size are determined, giving us a framework to analyse how these yield components relate to yield. Such a framework or conceptual model encompassing source and sink is a powerful analytical tool that provides in-depth insight into the production of yield at the seed and community levels. Our conceptual model does not, however, encompass information from the process, cellular, or organelle level, but, given the current state of our knowledge, this is probably not a significant disadvantage.

Our mechanistic description of the seed sink must include the two basic sources of variation in yield – environmental and genetic. Increasing yield is a matter of manipulating these two components; crop management manipulates the micro-environment to make it more favourable for plant growth and plant breeders modify the plant to improve its capacity to exploit the environment. Knowledge of the processes involved in the production of yield is not a prerequisite for increasing yield by either process (Evans, 1993, p. 266), but perhaps 'we can attain these ends more rapidly and more surely, however, if our experiments are guided by a higher level of understanding' (Duncan, 1969). We will use the concepts developed in Chapters 3 and 4 to evaluate the environmental and genetic aspects of the components of yield and how they relate to the yield of grain crops with the hope of approaching this 'higher level of understanding'. We will use Murata's (1969) division of yield production into the establishment of the yield container and filling the yield container to guide our discussion.

Size of the Yield Container

The maximum size of the yield container is determined by the number of seeds per unit area and potential seed size. Seed number is the yield component that accounts for most of the environmental variation in yield (see Figs 4.2–4.4). This position of primacy is the result of the simple fact that this component is determined, first, during the sequence of yield production and, second, because most crop species (with the exception of maize) have the capacity to make large adjustments in this component, i.e. it exhibits a great deal of plasticity. By virtue of its position at the beginning of the yield production process, the determination of seed number represents the first opportunity for the plant to adjust its reproductive output to environmental conditions. Structures responsible for its plasticity vary widely among species and were discussed in detail in Chapter 4.

Potential seed size determines the maximum size of the yield container once seed number is fixed. It is not clear how much the environment affects potential seed size or how often it limits yield, but, theoretically, it can be a limitation. Actual seed size is a representation of how well the yield container is filled. Seed size is determined after seed number, so it can only adjust to changes in the environment and assimilate availability after seed number is fixed. Consequently, environmentally induced variation in seed size is usually less than seed number (Figs 4.2, 4.3).

Duncan (1975) argued that:

> 'it is vital to any understanding of maize grain yield to know more about the physiological processes that determine ear and kernel number. Unfortunately, there is little published data known to the author, so we are faced with a subject too important to neglect but about which little experimental information is available.'

Duncan's statement applied equally well to other grain crops in 1975, but now our understanding of the determination of seed number is much more advanced. Mechanistic descriptions of the relationship between photosynthesis, partitioning, sink characteristics, and seed number have been developed and were discussed in Chapter 4. We can now evaluate with some confidence how to manage our crops to maximize seed number and yield.

Canopy photosynthesis

As described in Chapter 4, seed number is determined by the availability of assimilate from canopy photosynthesis during the critical period (Murata's (1969) Stage Two) when seed number is determined. The adjustment process is a matter of reducing potential seed number to a level that is in balance with the assimilate supply, assuming that potential seed number is greater than actual seed number, as it usually is, and that stress (e.g. high or low temperatures, nutrient deficiencies, moisture stress, lack of synchrony between pollen shed and silk appearance in maize) does not interfere with pollination. Thus, seed number and ultimately yield are directly related to the primary productivity of the plant canopy. Or, to put it another way, seed number in grain crops is usually source-limited, even in those crops (e.g. wheat) that have a reputation for being sink-limited. This observation is supported by a large body of literature demonstrating that changes in photosynthesis during the critical period cause corresponding changes in seed number as discussed in Chapter 4.

The identification of canopy photosynthesis as the primary determinant of seed number, and therefore yield, provides a mechanism for plant characteristics, environmental conditions and crop management practices to affect seed number. Maximum seed number for any cultivar of any species will occur only when the environment (above and below ground) is suitable for maximum canopy photosynthesis during the critical period. Meeting this goal must be a major objective of cropping systems designed to produce maximum yield.

Canopy photosynthesis is related to the photosynthetic characteristics of the plant, so species with C_4 photosynthesis should produce more seeds than species with C_3 photosynthesis, with all other factors, especially seed size and seed growth rate, held constant. Canopy photosynthesis is also influenced by environmental conditions, so the goal of crop management must be to create an optimum environment that minimizes stress and maximizes canopy photosynthesis during the critical period of seed number determination. Judicious selection of planting date and cultivar maturity may provide an opportunity to put the critical period in the most favourable environment for photosynthesis. The success of the Early Soybean Production System (Heatherly, 1999), which uses early maturing cultivars planted early to avoid stress in the mid-south of the USA, is a striking example of the potential of this approach.

Canopy photosynthesis and seed number reach maximum levels only when interception of solar radiation by the plant canopy is $\geq 95\%$ by the beginning of the critical period. Reaching this goal was greatly facilitated by the development of effective herbicides in many crops that eliminated the need for wide rows and mechanical cultivation, thereby allowing row spacing to decrease to ensure maximum radiation interception before the beginning of the critical period.

The need to reach maximum solar radiation interception by the beginning of reproductive growth is well documented, but does reaching 95% interception before the beginning of seed number determination provide any advantage? Early canopy closure would maximize crop growth rate earlier in vegetative growth, resulting in larger plants at the beginning of reproductive growth and increased total carbon capture during the crop's life cycle. These larger plants would not necessarily result in a higher crop growth rate during the critical period and, therefore, would not contribute directly to greater seed numbers.

Early canopy closure could, however, have indirect effects, both positive and negative, on seed number and yield. Early closure would reduce solar radiation levels below the crop canopy, reducing weed growth and competition and potentially increasing canopy photosynthesis. It could also reduce the number of herbicide applications needed for satisfactory weed control. Since water use by the crop (evapotranspiration) is related to leaf area, early closure could increase water use, resulting in possible stress during later growth stages, especially in those environments where water deficiencies frequently occur during reproductive growth.

In spite of the well documented relationship between seed number and canopy photosynthesis, some researchers still equate reproductive failure with lost yield, suggesting that simply decreasing reproductive failure and increasing seed number would increase yield. Reproductive failure occurs in most crops, as discussed in Chapter 4, as seed number adjusts to the availability of assimilate. Consequently, any increase in seed number from an artificial decrease in repro-ductive failure would, barring the unlikely stimulation of photosynthesis by the increased seed number, simply result in distributing the same amount of assimilate over a larger number of seeds, with no change in the final total product (Sinclair and Jamieson, 2008).

The well documented linkage between the assimilate supply during Murata's (1969) Stage Two and seed number, the primary determinant of sink size, iden-tifies canopy photosynthesis as the major determinant of seed number and yield. To argue that photosynthesis is not important as a yield determinant ignores the basic fact that almost all of the biomass accumulated by the crop comes from photosynthesis. High yields require high rates of canopy photosynthesis during reproductive growth. This link was described by C.M. Donald (see Donald, 1962) and it enjoys strong experimental support for all grain crops, but the idea that photosynthesis is not important is still prevalent. In recognition of the importance of photosynthesis, most modern high-input, high-yield cropping systems have evolved to maximize it and to minimize stress during reproductive growth to the maximum degree possible.

Length of Murata's Stage Two

Some researchers have suggested that seed number may be related to the length of Stage Two. The length of Stage Two varies among species and probably among environments, especially as determined by temperature, but whether or not this variability relates to seed number or yield is not always clear. There are at least two potential benefits associated with a long critical period. First, seed number could be directly related to the length of the critical period. Simply providing more time for flowering and seed set could increase seed number. The cumulative inter-cepted radiation and, therefore, the cumulative total assimilate production during the period, would increase in step with length, so if seed number is related to the total available assimilate during this period, length would be important.

A second potential benefit of a longer critical period is a reduction in the ef-fect of short-term fluctuations in the assimilate supply on seed number (Shibles *et al.*, 1975). A longer critical period may allow more time for compensatory ad-justments of seed number to a changing environment, resulting in greater sta-bility of seed number and reducing the effect of short-term stress events on seed number and yield. A shorter critical period would increase the likelihood that stress could last for the entire period, causing large reductions in seed number.

Is there evidence that a longer critical period, with its greater total assimi-late production, leads to a larger number of seeds? The photothermal quotient incorporated a length component when intercepted solar radiation was adjusted for temperature and the adjustment improved predictions of seed number from intercepted radiation (see Chapter 4) (Fischer, 1985; Ortiz-Monsataio *et al.*, 1994 – wheat; Islam and Morison, 1992 – rice; Cantagallo *et al.*, 1997 – sunflower; Poggio *et al.*, 2005 – pea), supporting the contention that length is important. Kantolic and Slafer (2001) lengthened the critical period in soybean in the field by manipulating photoperiod after initial bloom, and increased fruit and seed number and yield. Egli and Bruening (2000) reported a positive relationship between the length of the critical period (R1 to R5) and seed number, in comparisons involving several soybean cultivars and two planting dates. The linear association with length was stronger ($r^2 = 0.56$***) than with crop growth rate ($r^2 = 0.30$NS). The duration of the critical period has also been related to seed number in wheat (Slafer *et al.*, 2009). Fischer's (2011) formulation of the processes determining seed number in wheat included the duration of spike growth. These reports for several crops are consistent in suggesting that the length of the critical period is an important determinant of seed number.

There are also reports in the literature that the length of the critical period is not related to seed number. The length of the critical period of soybean (growth stage R1 to R5) increased by ~ 10 days as the total growth cycle increased from roughly 90–120 days across cultivars as maturity was delayed from maturity group (MG) 0 to MG IV (Egli, 1994a), and the longer critical period was associated with more nodes per plant (Egli, 1994a, 2013). However, there was little evidence that these changes *per se* increased seed number (Egli, 1993, 1997, 1999, 2013). A whole plant model of pod set of soybean (SOYPODP) predicted a reduction in pods per plant when the length of the critical period at each node and for the entire plant increased (Fig. 4.9) (Egli, 2015a).

The length of Stage Two can affect the temporal variation of flower production during the critical period. Flowers are usually produced over a longer period in species with longer critical periods than species with shorter critical periods. Eliminating this temporal variation by simultaneously pollinating all of the silks on the ear increased seed set of maize (Freier *et al.*, 1984; Carcova *et al.*, 2000). Increasing synchronous flower development in soybean without increasing the assimilate supply, by allowing one leaf to supply assimilate to the developing pods at three nodes, with a girdled node system, also increased seed set (Egli and Bruening, 2002). Abortion of late-developing flowers is often much higher than early-developing flowers (soybean, Heitholt *et al.*, 1986; maize, Otegui and Andrade, 2000). The second ear on maize develops after the first ear and usually does not produce kernels in low-photosynthesis environments (Vega *et al.*, 2001). Simulating increases in the number of early developing flowers for soybean with the pod set model SOYPODP (Egli, 2015a) also increased seed set, with no change in assimilate availability. These relationships can be explained as the result of competition for scarce assimilate with early developing or simultaneously developing fruits or seeds having an advantage over late developers (Egli, 2015a). The resulting decrease in the synchrony of flower production associated with a long critical period may reduce seed number.

There is evidence in the literature supporting an advantage for both long and short critical periods. It is possible, however, to argue against any advantage from a longer period from a theoretical viewpoint. Making length important seems to imply that the total assimilate accumulated during the critical period is important, but this is inconsistent with the notion that the yield production process is most efficient when seed number is matched to the ability of the canopy to fill the seeds. It can be argued that best approach to achieving this match is to relate seed number to the daily rate of canopy photosynthesis instead of the assimilate accumulated over a period of time. Involving the duration of the critical period would uncouple seed number from the growth rate, but filling the yield container depends directly on the rate of canopy photosynthesis during seed-filling. If seed number is related to length, a long period (high total assimilate accumulation) coupled with a moderate crop growth rate, for example, could produce more seeds than the moderate rate could fill, resulting in seeds that are smaller than normal. In spite of the data demonstrating value for a long critical period, it seems unlikely, at least in theory, that longer critical periods would increase yield.

Another potential benefit of a long critical period may be insulating seed number from short-term fluctuations in the assimilate supply. In theory, a long critical period should make it possible for a crop to recover from a short period of low assimilate availability and maintain seed number, whereas a short period would have less time (or no time at all if stress lasted for the entire period) to recover, and seed number could be substantially reduced, possibly resulting in a sink limitation during seed filling. From this viewpoint, the number of seeds resulting from a long critical period should exhibit a closer relationship to the average productivity (average assimilate availability) of the environment during a crop's critical period (Andrade *et al.*, 2005). In crops with shorter critical periods (e.g. maize or wheat), short-term stress could reduce seed number below the number supported by the average assimilate availability, thereby potentially reducing yield. Crop species with long critical periods with flowers produced throughout most of the period (e.g. soybean) should show more stability in seed number among environments than crops with shorter periods, but it is difficult to find data clearly supporting this advantage of a long critical period.

Seed number in soybean could not recover from 14 days of 60% shade at the beginning of the critical period, even though the shade ended at growth stage R3, roughly 30 days before pod production stopped (Egli, 2010). Similar results were reported in other field (Jiang and Egli, 1995) and greenhouse (Egli and Bruening, 2005) experiments. Pod survival after removal of the shade was not increased enough to overcome the loss during the shade treatment (Egli and Bruening, 2005), so the long flowering period did not eliminate the effect of these short stress treatments during the critical period.

Comparison of year-to-year variation of soybean and maize yield from long-term crop-rotation studies, however, suggests that soybean yield was less variable than maize (Table 5.1). The coefficient of variation for yield across years was larger for maize than soybean in 10 of 13 comparisons (average of 51% larger). Perhaps the longer critical period of soybean contributed to this greater yield stability. Stability of yield, however, does not necessarily imply equal stability of seed number, because variation in seed number can be offset to a degree by changes in seed size, making yield more stable than seed number. Species with

Table 5.1. Variation of maize and soybean yield in long-term tillage and rotation experiments.

		Continuous cropping				Rotation			
		Soybean		Maize		Soybean		Maize	
Location	Years (No.)	Yield (g m⁻²)	CV[1] (%)	Yield (g m⁻²)	CV[1] (%)	Yield (g m⁻²)	CV[1] (%)	Yield (g m⁻²)	CV[1] (%)
Lamberton, MN[2]	10	235	21.8	721	24.3	274	17.3	815	24.3
Wasceca, MN[2]	10	247	19.3	815	30.3	273	19.3	890	26.9
Arlington, WI[2]	10	351	20.9	847	27.7	355	20.9	947	20.8
West Lafayette, IN[3]									
Autumn plough	20	323	13.4	1072	16.0	352	11.2	1118	14.7
No-till	20	307	11.6	919	18.0	326	15.0	1087	13.2
Burlington, IA[4]									
Conventional-till	8	–	–	–	–	299	10.8	905	29.9
No-till	8	–	–	–	–	282	11.9	862	26.3
Boone Co., IA[5]	10	–	–	–	–	237	20.1	741	20.1

[1]Coefficient of variation.
[2]Adapted from Porter *et al.* (1998), 1986 to 1995 at each location from autumn plough treatment.
[3]Adapted from West *et al.* (1996), 1975 to 1994.
[4]Adapted from Brown *et al.* (1989), 1980 to 1987.
[5]Adapted from Karlen *et al.* (1995), 1984 to 1993. Conventional management (mouldboard ploughing) treatment.

more flexibility in seed size (e.g. legumes) may show a greater stability of yield than species with less flexibility in seed size (e.g. maize) (Andrade *et al.*, 1996; Borras *et al.*, 2004), even when the variation in seed number is not related to the length of the critical period.

Species comparisons of seed number stability are hampered by a lack of information on the length of the flowering period. Descriptions of the length of Stage Two usually ignore the actual period of flower production or pollination; a more refined estimate might reduce the supposed differences in length among species, which could explain the variation, or lack thereof, in stability. For example, soybean, usually considered to have a long critical period, produced 84% of its surviving fruits in less than 40% of its critical period (Egli and Bruening, 2006a), so does it have a long or a short critical period?

Stability in the face of short-term variation in assimilate supply could also be influenced by the relationship between assimilate supply and the survival of a fruit or seed. Stability would be enhanced by a delayed response of the reproductive structure to reductions in canopy photosynthesis and the supply of assimilate. Soybean fruits had to be exposed to low assimilate supplies for up to 16 days before they aborted (Egli and Bruening, 2006b), while 4–9-day shade treatments (60 or 80% shade) during peak pod production had no effect on seed number (Egli,

2010). Such a delayed response would limit the effects of short-term reductions in assimilate supply and perhaps mitigate any value of a long critical period. Other crop species probably exhibit similar responses. Large short-term changes in environmental conditions that effect photosynthesis may be relatively rare in many environments where grain crops are grown, further reducing the potential value of a long critical period.

In spite of suggestions that the length of Stage Two is important, the available evidence does not provide unequivocal support for the value of a long period for determination of seed number or for its stability. Relating seed number to the average productivity during the critical period would theoretically provide the best opportunity for the crop to produce high yield and a normal-sized seed, suggesting that a long critical period would have no value. Plants seem to have the ability to mitigate short-term fluctuations (fluctuations that are short compared to the length of Stage Two) in assimilate supply, diminishing any potential stabilizing value of a long critical period. I believe that the relative stability of seed size among environments of all grain crops (Figs 4.2 and 4.3), regardless of the characteristics of growth and reproductive development, suggests that crop plants have evolved a very efficient system of adjusting seed number to the productivity of the environment, and this system is probably not directly dependent upon the length of Murata's Stage Two.

Partitioning

In Chapter 4 we related seed number to the assimilate supply from photosynthesis, but only a fraction of the assimilate produced in a day is allocated or partitioned to reproductive growth. Consequently, descriptions of the yield production process, including the determination of seed number, commonly include a partitioning factor (Donald, 1968; Charles-Edwards et al., 1986, p. 24; see the discussion in Chapter 4 and Fischer, 2011), recognizing that seeds are only part of the dry matter produced by the crop. Increasing the proportion of assimilate allocated to reproductive growth during the critical period would increase seed number, but its potential affect on yield is less well defined.

Assimilate partitioning to reproductive growth during Murata's (1969) Stage Two will always be significantly less that 100%, given the need to sustain growth processes in the vegetative plant. Competing sinks for assimilate include synthesis of new leaf, stem and root tissues, growth and maintenance respiration, production of reserve materials and acquisition of N and other mineral nutrients. In some crop species, development of a terminal inflorescence ends vegetative development at the beginning of reproductive growth, which should make more assimilate available to set seeds; however, in other species with less determinate growth habits, vegetative growth may continue throughout part or all of the critical period. For example, node and leaf production of modern soybean cultivars continue throughout the critical period, usually ending near the beginning of seed filling (growth stage R5) (Egli and Leggett, 1973; Zeiher et al., 1982; Egli *et al.*, 1985a), which is after the time when most fruits begin growth (Egli and Bruening, 2006a). Increases in starch concentration in soybean leaves (Egli *et al.*, 1980) and nonstructural carbohydrates in maize (Uhart and Andrade, 1995b) and wheat stems (Evans *et al.*, 1975) during the critical period would also seem to reduce assimilate availability for seed set. It is difficult to estimate the degree to which these alternate sinks reduce seed number and whether these reductions limit yield. The picture is clearer in wheat, where the partitioning of assimilate to spike growth before anthesis is directly related to floret production, and eventually to seed number (Fischer, 2011). However, it is generally believed that vegetative growth should stop early in reproductive growth to maximize partitioning, seed number and yield; there is, unfortunately, little direct evidence supporting this proposition for most crops.

Although partitioning is important, it is nearly impossible to measure and the mechanisms that regulate it are poorly understood. This lack of understanding is intensified by our inability to describe mechanistically the exact link between the assimilate supply and reproductive survival. A number of mechanisms have been proposed (see Chapter 4 and Ruan *et al.*, 2012; Weber *et al.*, 1998; Boyer and McLaughlin, 2007), but none are well enough developed to help quantify the role of partitioning in determining seed number. We don't know, for example, whether stopping vegetative growth before the end of the critical period would increase seed number and yield, or how much the environment or management practices influence partitioning. To paraphrase Duncan (1975), we are left with a parameter that we know nothing about but is too important to ignore. Perhaps future research will uncover the mechanisms regulating partitioning and those linking seed number to the assimilate supply, allowing us to clearly evaluate the effect of competing sinks during the critical period on seed number.

Characteristics of the seed

Seed number is inversely related to genetic variation in SGR, as discussed in Chapter 4. This inverse relationship is predicted by the Charles-Edwards equation (equation 4.12 in Chapter 4), if we assume that the minimum assimilate flux requirement (a_G) of an individual seed is related to SGR. This relationship makes it clear that seed number is not just a function of the ability of the plant community to fix carbon, but the ability of the individual seed to utilize carbon is also important. The Charles-Edwards equation (equation 4.12) also makes it clear that variation in seed number as a result of variation in SGR (a_G) is not related to yield. Increasing SGR (a_G) at a constant level of assimilate availability causes a corresponding decrease in seed number (N_G). Genetic variation in SGR, environmental effects on basic seed characteristics (e.g. cell numbers) or composition-mediated effects on assimilate requirements for growth of an individual seed contribute to this inverse relationship.

There are many examples of this inverse relationship in the literature, including direct comparisons of cultivars with high and low seed growth rates (Fig. 4.6;

Egli, 1993) and species with a range in seed sizes (and presumably SGR) (Charles-Edwards *et al.*, 1986, p. 125). The oft-reported failure to increase yield when selecting for large seeds (Hartwig and Edwards, 1970), the classic case of yield component compensation (increase in seed size associated with decrease in seed number, with yield remaining constant), is a result of this inverse relationship. This relationship also explains the large variation in seed number among crop species (Table 4.2) that is not related to the capacity of the crop to produce assimilate.

The relationship between seed number and SGR provides a physiologically based mechanism that answers one of the major questions of yield component compensation, why genetically large seeds do not automatically produce higher yield. Many would argue that sinks that are highly active (i.e. have a high SGR) or are potentially large are exactly what is needed for high yield. Large and fast are adjectives frequently associated with maximum levels of production, for seeds; however, fast (i.e. high SGR) and large (where large results from fast) are completely yield neutral. Seeds that are large because they grow longer, however, may be associated with higher yield. Unfortunately, the large seed–high SGR combination is much more common than the large seed–long SFD pairing (see Chapter 3).

Summary

The size of the yield container (the number of seed and the potential seed size) is determined during Murata's (1969) Stage Two. Canopy photosynthesis and the availability of assimilate during this stage is the primary determinant of the seed number component of the size of the yield container and perhaps of potential seed size. This period is widely recognized as a crucial component of the yield production process. Maximum canopy photosynthesis during this period is needed to produce maximum yield and stress will reduce seed number and yield if the plant cannot counteract the decrease in number with an increase in seed size. This period is relatively short, exposing crops to potentially catastrophic yield losses when stress occurs throughout the entire period.

Most crops have substantial flexibility to increase seed number by producing more fruit-bearing structures and flowers, or increasing fruit survival in environments that produce high canopy photosynthesis. An example of this flexibility is the obvious capacity that many crops have to produce enough seeds to accommodate record yields that may be more than twice those normally encountered in farmers' fields. Even wheat and other cereals that have a reputation for being sink-limited during seed filling can increase seed number in high-yield environments. Some crops have lost this flexibility during domestication (e.g. maize), but it can be restored by adjusting population density (Egli, 2015b).

Relating the size of the yield container to the productive capacity of the crop matches size with the capacity of the crop to fill the container. A perfect balance between the size of the container and the capacity to fill it will produce the maximum yield that the environment will support and a normal-sized seed. This perfect balance will avoid a sink limitation (container too small), or a container that is too large resulting in small seeds that may have diminished commercial value. The perfect balance occurs, however, only when the productivity of the crop is constant during Stages Two and Three. This consistency may not occur in the field, because of random variation in environmental conditions or stress events, or due to seasonal changes in the environment during reproductive growth.

Interestingly, there are a number of reports showing a decrease in seed size as seed number increases in wheat (Fischer *et al.*, 1977; Evans, 1993, p. 262; Acreche and Slafer, 2006), suggesting that seed number and the ability of the canopy to fill the seeds fall progressively out of balance as productivity and seed number increase. The decrease in seed size in these reports was not large enough to seriously diminish the positive association between yield and seed number. This decrease in seed size could be due to increased competition among seeds for assimilate as seed number increases (Acreche and Slafer, 2006), implying that seed number was set too high relative to the ability of the crop to fill the seeds. Acreche and Slafter (2006) concluded, however, that the reduction in seed size in wheat resulted from an increase in the proportion of seeds from locations on the plant that produced smaller seeds.

The importance of variation in the length of the critical period and the temporal production of flowers and partitioning during the critical period are much less well understood, but they could also, at least theoretically, affect the number of seeds without a change in canopy photosynthesis (i.e. increase seeds per unit assimilate). We must note, however, that increasing the size of the yield container without an increase in canopy photosynthesis to fill the container will not increase yield, unless the container is too small, i.e. there is a sink limitation.

One important implication of linking seed number and the size of the yield container to canopy photosynthesis is that an increase in photosynthesis throughout the crop's life cycle will increase seed number and yield in most species. Higher plant populations may have to be used to capture the increase in maize. Most crop management systems have evolved to achieve maximum photosynthesis by the time the crop begins reproductive development as a result of the link between canopy photosynthesis and sink size. Overall, there is little evidence to suggest that the yield container is consistently too small for most crops in the field, leading to the conclusion that yield is primarily source-limited in most crops.

Filling the Yield Container

Filling the yield container is the last and most important phase of the yield production process. At the beginning of this phase, no yield has been produced; the first two phases were simply preliminary activities preparing for the production of yield. The vegetative plant is in place to produce the raw materials that will become yield and the yield container (number of seeds and potential seed size) has been established; now it is time to start filling it. Filling the container depends upon the supply of assimilate from the mother plant and the capacity of the individual seed to utilize the assimilate, because seed number is now fixed and can no longer change. Final seed size is a reflection of how well the yield container was filled. If the yield container is too small to accommodate all assimilate produced during seed filling, seed size should equal potential size and yield would be sink-limited.

The yield container is filled by assimilate produced by canopy photosynthesis during seed filling and by the redistribution of stored carbohydrates and N compounds produced before the beginning of seed filling. It is interesting that during this important phase of growth when all yield is produced, leaf senescence causes a decline in canopy photosynthesis that may eventually approach zero at the end of seed filling (physiological maturity). When assimilate is needed to fuel seed growth and the production of yield, the machinery supplying it is gradually destroyed.

Leaf senescence, the series of events that results in the cellular disassembly in the leaf (Thomas and Stoddart, 1980), progressively reduces the productivity of the plant canopy during seed filling of all grain crops. The N released by the disassembly of the photosynthetic apparatus is exported to the seed where it can account for up to 50–100% of the seed N at maturity (soybean, Zeiher *et al.*, 1982; wheat, Heitholt *et al.*, 1990; sorghum, Borrell and Hammer, 2000; maize, Below *et al.*, 1981). The decline in photosynthesis often starts early in the seed-filling period. In some studies, the decline of an upper leaf began when only roughly 40% of seed filling was completed (soybean, Boon-Long *et al.*, 1983; Secor *et al.*, 1983; maize, Pearson *et al.*, 1984), at anthesis (Wolf *et al.*, 1988b), 23 days after pollination in maize (Crafts-Brander and Poneleit, 1992), or ten days after the beginning of seed filling (growth stage R5) in soybean (Crafts-Brandner and Egli, 1987). Canopy photosynthesis of soybean started down slightly before the beginning of seed filling (growth stage R5) in some experiments (Wells *et al.*, 1982), but roughly at the mid-point of seed filling in others (Christy and Porter, 1982), while in wheat it started down at anthesis (Gent, 1995). In irrigated sunflower, canopy photosynthesis started declining at the beginning of seed filling and was only 30% of the initial level at the mid-point of seed filling (Hall *et al.*, 1990). The rate of senescence apparently varies among genotypes (Sinclair, 1980) with 'stay green' types (identified in numerous species including maize (Rajcan and Tollenaar, 1999; Duvick, 2005) and sorghum (Borrell and Hammer, 2000)) representing the ultimate reduction in the rate of senescence. Soybean producing record high yield retained green leaves at maturity (Purcell, 2008). The rate of senescence is also accelerated by water or N stress (Aparico-Tejo and Boyer, 1983; Wolf *et al.*, 1988b; Hayati *et al.*, 1995; de Souza *et al.*, 1997; Brevedan *et al.*, 2003) and by some leaf diseases (Dimmcock and Goodin, 2002; Pepler *et al.*, 2005). The rate of leaf senescence must be sensitive to temperature; longer SFDs at lower temperatures must be associated with lower rates of senescence. Photosynthesis is the primary source of assimilate for seed filling; consequently, variation in the timing and rate of senescence could affect yield.

Canopy photosynthesis is not the only source of assimilate during seed filling: amino acids are released during senescence and carbohydrates accumulated in leaves and stems (sucrose, starch and fructans) during vegetative growth can be redistributed to the seeds. The contribution of these sources to yield is highly variable, with redistributed N accounting for up to 100% of the mature seed N in soybean (Egli *et al.*, 1978b; Zeiher *et al.*, 1982), while estimates for stored carbohydrates vary from 8% of the final yield for soybean (Egli, 1997), to 20–30% for wheat, barley, rice and sunflower (Foulkes *et al.*, 2009), and 0–7% for maize (Swank *et al.*, 1982). The contribution often increases when stress occurs during seed filling, but it is not always clear that this increased relative contribution equates to a reduction in the effect of stress on yield (Hall *et al.*, 1989).

Seed growth rate (SGR)

Seed number is fixed at the beginning of Stage Three, so the capacity of the individual seed to accumulate dry matter (SGR, SFD and potential seed size) plays an important role in filling the yield container. The capacity to accommodate the assimilate supplied by the vegetative plant is limited because the SGR eventually saturates as assimilate availability increases (see Chapter 3), so there is a limit to how much SGR (and, by extension, the total seed growth rate per unit ground area) can respond to increases in assimilate supply during seed filling. If seed number is set too low, SGR may not be able to increase to use all the assimilate, potentially limiting yield. Some data suggest that seeds are usually growing at or near their maximum rate in the field (see Chapter 3), so the capacity of the seed to respond to an increase in photosynthesis during seed filling may be limited unless the excess assimilate translates into a longer seed-filling period. Of course, there would be no limit of the response to a decrease in photosynthesis, a reduction in SGR, total seed growth rate and yield.

Seed-fill duration (SFD)

Filling the yield container also depends upon how long filling continues, i.e. the length of the seed-filling period, consequently, environmental or genetic variation in SFD is often related to yield with all other yield determining characteristics held constant. There are potential interactions between SFD and SGR that could modify the relationship between SFD and yield. For example, reductions in SFD at higher temperatures could be offset by higher SGR, so that seed size and yield are not affected (Chowdhury and Wardlaw, 1978), or increases in SGR when potential seed size limits final seed size will actually reduce the SFD (Kato, 1986). In most situations, however, the length of the seed-filling period is related to yield. Since seeds cannot grow without a source of assimilate, changes in SFD must be associated with changes in senescence patterns during seed filling (Crafts-Brandner and Poneleit, 1992).

Positive associations between SFD and hybrid or cultivar yields were found in maize (Daynard and Kannenberg, 1976; Bolanos, 1995), wheat (Gebeyehou *et al.*, 1982; Penrose *et al.*, 1998), barley (Leon and Geisler, 1994; Dofing, 1997) and soybean (Hanway and Weber, 1971; Dunphy *et al.*, 1979). The higher yield of maize hybrids compared with inbreds was associated with a longer SFD (Johnson and Tanner, 1972; Poneleit and Egli, 1979). Selection for long seed-filling duration resulted in higher yields in soybean (Smith and Nelson, 1986a,b) and maize (Cross, 1975; Crosbie and Mock, 1981).

Selection for higher yield produced cultivars with longer seed-filling periods in oat (Helsel and Frey, 1978), groundnut (Duncan *et al.*, 1978), soybean (Gay *et al.*, 1980; McBlain and Hume, 1980; Boerma and Ashley, 1988), durum wheat (Motzo *et al.*, 2010), and maize (Russell, 1991). The increase in average maize yields in Indiana from 1950 to 1980 was associated with an increase in SFD (Fig. 3.7, McGarrahan and Dale, 1984). Increases in yield and seed size are associated with domestication of many crops (Harlan, 1992) and domesticates usually produce higher yields which could have been associated with lengthening the seed-filling period. In these examples, selecting for yield inadvertently lengthened the seed-filling period.

The plant's environment affects SFD and these effects are frequently translated into changes in yield. Water stress shortened seed filling and reduced yield of chickpea (Davies *et al.*, 1999), soybean (Meckel *et al.*, 1984; de Souza *et al.*, 1997), maize (NeSmith and Ritchie, 1992), barley (Aspinall, 1965), sunflower (Whitfield *et al.*, 1989) and wheat (Frederick and Camberato, 1995). Water-logging of wheat at jointing and/or anthesis accelerated senescence and shortened the seed-filling period (Araki *et al.*, 2012). A shorter seed-filling period played a role in yield reductions from N stress with wheat (Frederick and Camberato, 1995) and soybean (Egli *et al.*, 1978b), and P and K stress with maize (Peaslee *et al.*, 1971). Nitrogen stress reduced leaf area duration and yield of maize (Wolf *et al.*, 1988a), probably as a result of a shorter seed-filling period. The effects of sowing date and irrigation on yield of pinto beans (*Phaseolus vulgarius* L.) and field beans (*Vicia faba* L.) was expressed through changes in leaf area duration, which probably represented differences in SFD (Husain *et al.*, 1988; Dapaah *et al.*, 2000). Wheat yield was closely associated with leaf area duration across trials involving planting date, seeding rates and N fertilizer rates (Fischer and Kohn, 1966). Controlling leaf diseases with fungicides increased SFD in wheat (Dimmock and Gooding, 2002). The effect of stress on SFD and yield is often 'hidden' because the acceleration of senescence and the shorter seed-filling period is not obvious without a non-stressed control for comparison, so the stress may not be noticed until the crop is harvested.

Seed-fill duration is sensitive to temperature, and this variation frequently translates into changes in yield. Artificially lowering night temperature increased yield of wheat maize, and soybean, apparently as a result of a longer seed-filling period (Peters *et al.*, 1971). Lower temperatures and longer seed-filling periods increased yield of oat (Hellewell *et al.*, 1996), and wheat (Wardlaw *et al.*, 1980). Lower temperatures and longer SFDs contributed to larger maize yields at higher elevations (Cooper, 1979). Compensatory effects of solar radiation (Muchow *et al.*, 1990) or seed-growth rate (Chowdhury and Wardlaw, 1978), however, minimized changes in seed size and yield in other situations.

Long seed-filling periods may be partially responsible for exceptionally high yields in environments with moderate temperatures (Duncan *et al.*, 1973; Muchow *et al.*, 1990; Sinclair and Bai, 1997). Duncan *et al.* (1973) theorized that an environment with a high daytime temperature (maximum photosynthesis) and a low night temperature (slow development and longer SFD) might produce maximum yield, assuming high radiation and no other limiting factors. Muchow *et al.* (1990) found that exceptionally high maize yields (> 1600 g m^{-2}) occurred only at locations with high solar radiation and lower temperatures that resulted in longer growth durations, confirming this concept.

Genetic differences in SFD, however, are not always related to yield. Examples come from Wych *et al.* (1982) and Peltonen-Sainio (1991) with oat, Sexton *et al.* (1994) with bean, Dwyer *et al.* (1994) with early maturing maize hybrids and Van Sanford (1985) with soft red winter wheat. Genetic selection for long SFD in maize was successful, but SGR was reduced and yield not changed (Hartung *et al.*, 1989).

Filling the yield container involves both the total seed growth rate and the duration of seed dry matter accumulation, and either could limit yield. While it's true that total seed growth rate, through its association with seed number, probably accounts for more yield variation among crops and environments than SFD, SFD is also important, and ignoring it is a mistake, especially when we consider the challenges of increasing yield in the future.

So far, our discussion of the association between the duration of seed filling and yield has focused on long seed-filling periods and higher yield. We must, however, also be aware of the opposite aspect of SFD; short filling periods are a serious obstacle to high yield. For example, filling periods of some cultivars of grain legumes (e.g. cowpea, Lush and Evans, 1981; Wien and Ackah, 1978, and common bean, Sexton *et al.*, 1994) are as short as 5–15 days; much shorter than 30–40 days for many cultivars of more widely grown crops, such as maize, soybean or wheat (Fig. 3.2). Since final yield is the product of total seed growth rate and SFD, a cultivar with a short seed-fill duration will need an exceptionally high total seed growth rate to produce high yield. Some have argued that short filling periods and high total seed growth rates are a way to increase yield in environments with short growing seasons (Whan *et al.*, 1996); however, the potential for this approach may be limited by the difficulties associated with increasing the rate of dry matter accumulation by the crop community and the total seed growth rate. Although a short-filling period restricts yield potential, it does allow the crop to complete its life cycle in a short time, making the crop more adaptable to environments with short growing seasons and more useful in multiple cropping systems. We will discuss involvement of life-cycle length in agricultural productivity later in this chapter, in the section on time.

The enigma

The processes involved in the filling of the yield container present an enigma – the accumulation of dry weight by the seeds (total seed growth rate per unit area) is best described by a linear function, while canopy photosynthesis declines steadily during seed filling, often approaching zero at physiological maturity. The constant total seed growth rate, ignoring, of course, the lag phases at the beginning and end of the seed-filling period, is well documented for most grain crop species (e.g. soybean, Koller *et al.*, 1970; Egli and Leggett, 1973; maize, Duncan, 1975; wheat, Fischer and Kohn, 1966).

The decline in canopy photosynthesis during seed filling is well documented and occurs in all crops, although the temporal characteristics of the decline are variable among and within species, and environments (see previous discussion of senescence earlier in this chapter and in Chapter 3). The decline often begins relatively early in the seed-filling period, and canopy photosynthesis can approach zero roughly when the seed filling stops at physiological maturity. Consequently, the supply of assimilate from photosynthesis is usually declining during seed filling in all grain crops. How does this steady decline in source activity support a constant rate of dry matter accumulation in the seeds? I think there are at least three potential explanations for this phenomenon.

First, the redistribution of stored carbohydrates and N from vegetative tissues to the seed could make up for the decline in photosynthetic activity and produce the linear seed growth curve. Redistribution occurs in most crops (see the previous discussion in this chapter) and provides another source of assimilate to supplement that coming from canopy photosynthesis. The contribution of redistributed assimilate to yield varies among grain crop species and environments, but most estimates of the contribution are less than 30% of the final yield. Redistributed N can account for up to 100% of the seed N at maturity, but this is a relatively small proportion of the total seed weight for most species. Unfortunately, there is rarely enough data available for a quantitative evaluation of the ability of redistribution to make up for the 'missing' assimilate.

Changes in partitioning during seed filling that make a greater proportion of the assimilate from photosynthesis available for seed growth is a second potential mechanism to match a declining rate of canopy photosynthesis with a constant total seed growth rate. All the assimilate produced by photosynthesis is never available for reproductive growth, although we often tend to assume that most of it goes to reproductive plant parts when the production of leaves, stems or other vegetative plant parts stop. Vegetative tissues, however, require assimilate for maintenance respiration and other growth processes, such as ion uptake and N acquisition. The assimilate partitioned to these processes may decrease as the leaves senesce during seed filling, thereby increasing the amount available for seed growth (Tanaka, 1980, as cited by Fageria *et al.*, 2006, p. 126). Detailed quantitative estimates of the magnitude of these potential shifts in partitioning are not readily available, but it seems likely that increased partitioning to the seed late in seed filling could make a significant contribution to maintaining the linear total seed growth rate as the source activity decreases.

Finally, perhaps our conclusion that total seed growth follows a linear curve is not correct. A steady decrease in the total seed growth rate during seed filling would be consistent with the declining canopy photosynthesis rate. A linear curve usually provides an adequate fit to total seed dry weight during seed filling, but, given the variability often associated with dry weight measurements in the field, it's possible that statistical analysis did not detect the true curvilinear nature of the curve.

Which of these three options best explains the constant seed growth rate throughout seed filling? The argument that the linear total seed growth rate is, in fact, curvilinear is, in my opinion, the least likely explanation. The linear growth curves for individual seeds reported for many crop species (see Chapter 3), combined with a stable seed number during seed filling would produce a constant total seed growth rate during most of the filling period. It is hard to imagine that a steadily declining total seed growth rate would remain undetected.

There is, I believe, enough indirect evidence to suggest that the combination of remobilization of storage reserves from vegetative plant parts and a decline in partitioning of assimilate to the vegetative plant are responsible for maintaining the linearity of total seed growth curve. The temporal decline in photosynthesis during seed filling varies among species and environments, as does the contribution of redistributed assimilate; one wonders how these disparate processes are coordinated to maintain the constant rate of seed growth for all grain crop species in environments representing a wide range in productivity.

Seed size and yield

Seed size provides a very visual representation of yield; a representation that is more conspicuous than seed number. Because seed size is such an easily observable characteristic and because of the large variation in size among and within species (see Table 3.1 and the discussion in Chapter 3), it attracted the interest of early plant breeders, agronomists and crop physiologists. It must have seemed obvious that large seeds would equate to high yield; however, as we have seen previously, this relationship can be misleading, because seed size may or may not be related to yield. Large seeds are not always an indication of high yield and small seeds are not a reliable indicator of low yield. Seed size is under genetic control and it is influenced by environmental conditions during seed filling (Murata's (1969) Stage Three); whether or not seed size is related to yield depends upon the source of variation (genetic or environmental) and the mechanism responsible for the variation in seed size

Most of the genetic differences in seed size are a result of variation in SGR, which results in compensatory changes in seed number with no effect on yield. This compensation was discussed previously in Chapter 4 and has been documented experimentally for many crops. Examples include comparisons involving multiple cultivars of common bean (White, 1981; Sexton *et al.*, 1994), soybean (Table 4.1), maize hybrids and maize inbreds (Poneleit and Egli, 1979), and barley (Hamid and Grafius, 1978). Such relationships probably exist for all grain crops. Breeding for large seeds usually does not increase yield, as Hartwig and Edwards (1970) discovered when they developed soybean isolines with large differences in seed size, but no difference in yield.

The seed number–seed size compensation is even more obvious when comparing crop species where the variation in seed size is much larger. Both large and small seeds are associated with high average yields in Table 4.2. Rice produces relatively high average yield with small seeds (28 mg seed^{-1}), while soybean and common bean produce very modest yields with relatively large seeds (202–345 mg seed^{-1}). Most of the variation in size among species is related to SGR (e.g. rice seeds grow much more slowly than soybean or common bean) (Table 3.1) and, as discussed previously, seed number adjusts to the differences in SGR, thus eliminating any effect on yield.

Genetic variation in size that is related to SFD, however, is related to yield. Seeds that are large because they grow for a long time will result in higher yield. Genetic differences in size that are related to SFD are rare, but they do exist. Our analysis of genetic differences in seed size indicates that it would be a mistake to judge the yield potential of a cultivar or a crop by the size of its seeds.

Environmental variation in seed size is an indication of how well the crop community can fill the yield container within the limits set by potential seed size. The response of seed size to increases in assimilate supply during seed filling may, however, be limited by the capacity of SGR to respond to the additional assimilate (i.e. sucrose concentration in the seed may already be high enough to saturate SGR; Fig. 3.3) or the potential to increase SFD. Both of these responses can be limited by potential seed size (the maximum seed size that cannot be exceeded). Stress during seed filling that reduces photosynthesis or accelerates senescence will always cause reductions in seed size and yield.

Interactions between SGR, SFD and potential seed size determine the response of seed size to changes in the assimilate supply during seed filling. Limitations by potential seed size may be more common in some crop species than in others (e.g. legumes vs cereals) (Borras *et al.*, 2004) and species with greater flexibility are better situated to capitalize on increased assimilate supplies during seed filling by increasing seed size and yield. In contrast, there are no limitations to reductions in seed size and yield from decreases in the assimilate supply during seed filling; consequently, small seeds and lower yields are probably more likely in the field than larger seeds and higher yield.

In summary, the answer to the question: Does seed size relate to yield? is both Yes and No, a confusing answer on the surface, but one that can be easily understood by considering the components of seed size and the source of variation. The 'no' answer relates to size variation caused by genetic variation in SGR, the most common cause of seed size variation within and among crop species. The 'yes' answer takes into account genetic differences in size that are a result of variation in SFD and size variation associated with changes in the assimilate supply during seed filling. These principles apply to all crop species, although the variability in and the importance of seed size as a yield component probably varies among them. Interestingly, the substantial genetic variation in seed size that is determined by SGR is relatively easy to manipulate genetically, but it is not related to yield. The dream that yield could be increased by simply increasing seed size, an idea still occasionally touted by biotechnologists (see, for example, Ma *et al.*, 2015), is just a dream that cannot be translated into reality. The capacity of the seed to respond to variation in the assimilate supply during seed filling is an important component of the evaluation of source–sink limitations, which will be discussed in the next section.

Source–Sink Limitations of Yield

What limits yield, the ability of the plant community to produce assimilate (the source) or the ability of the seeds to utilize assimilate produced by the leaves (the sink)? The separation of the production of yield into sources and sinks has long been a popular approach among crop physiologists. In the general sense, a source is a plant part producing assimilate, usually via photosynthesis, although translocation of reserve assimilate from a plant part could qualify that part as a source. A sink is a plant part importing and utilizing assimilates from the source. More precise and complex definitions could probably be constructed, but these will serve our purpose.

From the viewpoint of our interest in the seeds of grain crops, leaves are the source and seeds are the sinks. Considering only leaves and seeds is an obvious simplification because there are other photosynthetic organs (stems, awns, carpels, etc.) and other sinks (new leaves, active growing points, nodules, roots, etc.), but this simplified system is usually used by crop physiologists and is adequate for our evaluation of source–sink limitations. Yield is basically a function of the production of assimilates by the leaves in the plant canopy and the utilization of these assimilates to synthesize reserve materials in the seeds, making it possible to visualize a limitation by either source or sink.

This question of source vs sink limitation is important because of the implications for yield improvement. If the source is limiting, yield improvement efforts should focus on the source and attempts must be made to increase the plant's assimilatory capacity. But if the sink is limiting, focusing on the source is foolish, and attention must be given to improving the size of the sink or its ability to accumulate complex carbohydrates, oil and protein. Historically, this question was not as relevant as it is today; plant breeders in the past increased yield very successfully by selecting for yield *per se* and not worrying about whether they were changing the source or the sink. Now that it is easier to focus plant improvement efforts on individual growth processes, the source–sink question is more important.

Claims of either a source or sink limitation for many crops can be found in the literature (see, for example, Fageria *et al.*, 2006, p. 117; Borras *et al.*, 2004; Borras and Gambin, 2010). Evans (1993, pp. 172–185), after a thorough review of the subject, concluded that source and sink are not independent and therefore both may limit yield. The lack of independence stems from the effect of the source on the size of the sink and the hypothesized ability of the sink to influence the activity of the source. Concluding that both may be limiting is not very satisfying and does not provide much guidance for future yield improvement efforts.

It is important when considering source–sink limitations to remember the sequential nature of the yield production process (discussed in Chapter 4). First, the size of the yield container is established (the number of seeds and potential seed size) and then the yield container is filled. Consequently, when evaluating source–sink limitations, the growth stage when alteration treatments are applied must be considered when interpreting the results. Ignoring this sequence may lead to erroneous conclusions and, in particular, an over-estimation of the importance of sink limitations. It is important to remember that seed number, which is the major determinant of the potential size of the yield container, is determined first, followed by seed size.

Our considerations of the determination of seed number clearly describe the seed number component of yield as source limited. The Charles-Edwards equation (equation 4.12, Charles-Edwards *et al.*, 1986, pp. 124–127) relates seed number directly to canopy photosynthesis during the critical period for seed number determination (Murata's (1969) Stage Two). Evidence supporting this relationship, discussed previously, includes the well-documented association be-tween estimates of canopy photosynthesis and seed number in many crop species. Manipulation of photosynthesis during this period (shade, CO_2 enrichment, de-foliation, or solar radiation enrichment) usually results in a corresponding change in seed number. Environmental conditions and crop management practices often affect yield by causing variation in canopy photosynthesis and crop growth rate and, ultimately, seed number. The 'Golden Rule' of crop physiology – maximum yield requires maximum solar radiation interception early in reproductive growth – is based on source control of seed number and yield.

Artificially increasing the size of the sink (number of seeds) usually does not increase yield, supporting the contention of a source limitation. The auxin transport inhibitor TIBA (tri-iodobenzoic acid), sprayed on soybean, often in-creased seed number, but seed size decreased and yield did not change (Tanner and Ahmed, 1974). Use of a multiple ovary or a large spike trait to increase seed number in wheat did not increase yield (Gaju *et al.*, 2009). Carbon dioxide enrich-ment treatments only during the critical period for seed number determination in wheat (Fischer and Aguilar, 1976) and soybean (Hardman and Brun, 1971) significantly increased seed number and that translated into higher yield in wheat, but not soybean. Walker *et al.* (1988) increased seed number relative to the size of the source by moving pots containing maize plants closer together during seed filling, but the larger sink had no effect on yield. Yield of most crops does not re-spond to simply increasing sink size without increasing the capacity of the source to fill the sink. The exceptions in the literature may be a result of the difficulties associated with experimentally manipulating sink size without influencing source activity during seed filling. These exceptions need not negate our conclusion that sink size is usually source-limited.

An obvious exception to source control of seed number is the special case where stress-induced failure of pollination or fertilization (high or low temperat-ures, boron deficiency, moisture stress (Satake and Yoshida, 1978; Herrero and Johnson, 1980; Warrag and Hall, 1984; Bolanos and Edmeades, 1996; Rawson,

1996)) reduces seed number. The assimilate supply would then exceed the potential seed number (i.e., flower number) and yield would be sink-limited. Sink limitations can also occur when maize is grown at populations that are too low to produce enough flowers to accommodate the available assimilate (Egli, 2015b). In these situations, seed number is not related to the assimilate supply and yield is limited by the size of the sink. These sink limitations represent exceptions to the general rule and they are relatively rare occurrences, but they can have catastrophic effects on yield. Ignoring these exceptions, seed number and thereby yield are source-limited in all grain crops; increasing canopy photosynthesis throughout reproductive growth will increase yield.

After seed number is fixed, yield is determined by processes involved in filling the yield container (i.e. SGR, SFD (equation 4.13) and potential seed size). If the assimilate supply during seed filling exceeds the capacity of the sink to utilize it, the sink limits yield. If, on the other hand, the sink can accommodate an increase in assimilate supply, there is a source limitation. Experimental evaluation of source–sink limitations during seed filling usually involves artificially reducing the size of the sink (removing seeds) to increase the assimilate supply per seed or increasing the source with no change in sink size and then observing the response. Failure of the seeds to increase in size is taken as an indication of a sink limitation. The assumption that the treatments actually modified the supply of assimilate to the seed is rarely verified (Jenner, 1980).

Increasing the assimilate supply will increase seed size only if the seeds can grow faster or if they can grow for a longer time and potential seed size is not limiting. Increasing SGR depends upon the capacity of the seed to respond to an increase in assimilate supply. Since the response of SGR to assimilate supply follows a saturation curve (see Fig. 3.3), increasing assimilate supply per seed may or may not increase SGR, depending upon the effective assimilate or sucrose concentration in the seed. If an increase in the assimilate supply has no effect on SGR, i.e. the assimilate concentration is already on the saturation part of the curve, seed size may not increase. Of course, if the assimilate level in the seed is below the saturation level, increasing assimilate supply will increase SGR.

Seed size will also increase if SFD can increase to utilize the extra assimilate. Increasing SFD requires a delay in leaf senescence to provide assimilate for the extension of seed growth; once the senescence process is complete, there will be no more assimilate available to the seed and growth cannot continue.

The potential seed size component of the yield container can also limit seed size increases by creating an absolute barrier to increases in seed size. If potential seed size is limiting, a higher SGR in response to increases in assimilate supply, will result in a shorter seed-filling period and no change in seed size. The same argument can be made for SFD; it can increase only within the limits set by potential seed size.

Whether or not a sink limitation exists during seed filling is determined by the ability of the sink to accommodate extra assimilate. If the extra assimilate can be accommodated, seed size will increase and a source limitation would exist. If not, seed size would not change and the crop would be sink-limited during seed filling.

From the viewpoint of the individual seed, extra assimilate can be accommodated by growing faster or longer, but either of these responses can be limited by potential seed size. The complexity of the response system, involving three somewhat independent seed growth characters, could account for the variety of responses among and within cultivars, species and environments obtained in experiments testing sink limitations during seed filling. Many of the possible responses described here have been documented in several crops.

For example, some experiments have shown no response of SGR when the source is enhanced relative to the sink, usually accomplished by reducing the sink by depodding or degraining (soybean, Egli *et al.*, 1985a; wheat, Slafer and Savin, 1994; maize, Frey, 1981), but other experiments have shown that SGR will respond to enhanced assimilate supply (soybean, Egli *et al.*, 1985a; wheat, Simmons *et al.*, 1982). Depodding soybean plants increased the SFD (Egli *et al.*, 1985a); however, degraining wheat did not result in a longer seed-filling period (Simmons *et al.*, 1982).

Similar conflicting responses were noted when the effect of sink-reduction treatments was evaluated by changes in seed size, without considering SGR or SFD. Sink-reduction treatments in wheat usually did not increase seed size (Slafer and Savin, 1994; Calderinni and Reynolds, 2000), but exceptions can be found (Ma *et al.*, 1995; Cruz-Aguado *et al.*, 1999). The situation is the same for maize, with both no response (Jones and Simmons, 1983, Gambin *et al.*, 2008) and increased seed size (Gambin *et al.*, 2008) reported. Seed size in soybean (Egli *et al.*, 1985b; Munier-Jolain *et al.*, 1998) and other grain legumes (Munier-Jolain *et al.*, 1998) generally responded to reductions in fruits or seeds. Sorghum was also very responsive (Gambin and Borras, 2007). Seeds of canola (*Brassica napus* L.) (Fortescue and Turner, 2007) and sunflower (Steer *et al.*, 1988) increased in size, while rice responded in only three of eight cultivars (Kato, 1986). Borras *et al.* (2004) found that soybean seed size was more responsive to sink reduction during seed filling than either maize or wheat.

Clearly, crops may or may not be sink-limited during seed filling, although it seems that some species may have a greater propensity to be sink-limited than others (e.g. rice or wheat vs soybean). The mechanisms responsible for failure of the yield container to accommodate the extra assimilate in some situations, but not in others, are rarely determined.

Our overall conclusion is similar to that reached by Evans (1993, pp. 184–188): both source and sink can limit yield. In the real world of the farmer's field, however, yield is primarily and predominantly source-limited. Seed number is the first yield component fixed in the sequential yield production process and it is determined by the assimilate supply (i.e. the source) during Stage Two of the yield production process. Since yield is primarily a function of how many seeds are produced, and seed number is source-limited, yield is primarily source-limited. High yields require high levels of canopy photosynthesis during reproductive growth and any increase in canopy photosynthesis during the entire reproductive growth period (flowering, seed set and seed filling) will result in an increase in yield. This relationship holds for all crop species, even those like wheat, for example, that have a reputation of being sink-limited. There is no evidence that crops grown with accepted management practices cannot increase seed number to match the productivity of the environment, up to levels associated with record yields.

There is also evidence that the sink can limit yield. The significance of this limitation in field production environments is not clearly understood, but its effect on yield is probably much less than the source limitation just discussed. The fact that seed size cannot increase when sink size is artificially reduced tells us very little about the magnitude of yield loss from this sink limitation, or the potential increase in yield if the sink limitation was eliminated. Since seed number adjusts to canopy photosynthesis and the assimilate supply, sink size and the ability to fill the container should be reasonably balanced. This balance could result in a sink that could accommodate all of the available assimilate, in spite of the failure of the seed to increase in size when sink size was artificially reduced. In this case, there would be no effective sink limitation in the field, so traditional source–sink evaluations maybe misleading. Yield loss would occur only if the seed could not accommodate all the assimilate available during seed filling.

The separation in time of the determination of seed number from seed filling allows changes in environmental conditions between the two phases to create an imbalance that could result in a sink limitation, either from short-term fluctuations in the environment (e.g. drought stress only during seed set reduces seed number) or from seasonal changes in solar radiation from Stages Two to Three. These seasonal changes would favour a sink limitation in winter-grown crops (solar radiation levels would increase from Stages Two to Three, but not in summer-grown crops (solar radiation would decrease from Stages Two to Three) (Egli, 1999; Egli and Bruening, 2001; Borras *et al.*, 2004). It is not possible to estimate how much these sink limitations reduce yield; it seems, however, that the effectiveness of the adjustment of seed number to the assimilate supply in most crops and the usual persistence of environmental conditions in field environments would minimize yield losses.

Although our general conclusion is that either the source or the sink can limit yield, the source limitation is, by far, the more important in all grain crops. Consequently, increasing the rate of canopy photosynthesis during reproductive growth will increase yield. It is a serious mistake to imply that simply increasing sink size will increase yield (Sinclair and Jamieson, 2008), the usual implication of a sink limitation. Emphasizing a sink limitation implies that photosynthesis is not an important yield determinant, which is also a serious mistake. The importance of photosynthesis is not surprising, given that the bulk of the biomass and yield comes from this process. When it comes to yield of grain crops – the source rules.

Partitioning and Harvest Index

The term 'partitioning' implies a division into separate parts, dividing the whole into fractions and, when used by crop physiologists, it usually refers to dividing assimilate from photosynthesis among the various plant parts and metabolic sinks that constitute the plant and plant growth. Assimilate is used to produce structural tissues (leaves, stems, petioles, roots, seeds) and to drive metabolic processes that produce and maintain these tissues (e.g. growth and maintenance respiration, N acquisition, etc.). Assimilate must be partitioned among these sinks in the appropriate proportions; proportions that change substantially during the crop's life cycle. The consistency of plant form and function among the many environments where crops are grown suggests a rather precise regulation of partitioning. Plants seldom partition too much assimilate to roots and not enough to leaves, or acquire more N than the crop needs. Relative stability of seed composition reflects stable partitioning of assimilate among starch, protein and oil. Partitioning is complex and dynamic, but from the perspective of yield production, crop physiologists usually focus on the partition between vegetative and reproductive growth, i.e. production of new vegetative tissues and maintenance of already existing tissues, and the production and growth of seeds.

Partitioning patterns obviously change during crop development. Partitioning to reproductive growth is zero before reproductive growth begins (ignoring the growth of the structures that will ultimately bear the reproductive plant parts) and increases to a maximum at some point during reproductive growth, after vegetative growth stops. The role of partitioning in determining seed number was discussed earlier in this chapter. Since assimilate from canopy photosynthesis drives plant growth and the production of yield, increasing partitioning of assimilate to reproductive plant parts is likely to increase yield.

We have divided the production of yield into three stages (see Chapter 4): vegetative growth; flowering and establishment of the yield container; and filling the yield container. These are, from a functional viewpoint, distinctly separate stages, but, on a developmental basis, there is frequently some overlap between vegetative growth and the establishment of the yield container. Partitioning between these two stages can be easily understood because they occur at the same time. Stages one and three, however, occur at different times, with vegetative growth generally complete before seed filling begins, making it difficult to understand how significant amounts of assimilate could be transferred or partitioned from one phase to another. Assimilate produced during vegetative growth (Stage One) cannot be readily shifted to seed filling (Stage Three), except by remobilization of stored assimilate. The separation in time of the various activities involved in the production of yield must be included in any mechanistic evaluation of partitioning.

The classic measure of partitioning widely used by crop physiologists is the harvest index first popularized by Donald in the 1960s (Donald, 1962, 1968), although it was, according to Evans (1993, p. 238), first used by Roberts in an 1847 comparison of wheat cultivars. The harvest index (HI) is the ratio of seed yield (SY) to the total biomass (vegetative mass (VM) + seed yield) (see equation 5.1) at maturity; consequently, it represents an estimate of partitioning when the yield production process is complete.

$$HI = SY(VM + SY)^{-1} \qquad\qquad (5.1)$$

Usually the vegetative mass includes only the above-ground portions of the plant, and the roots are ignored. The HI is easy to measure on many crops, requiring only the additional measurement of vegetative mass at maturity. In fact, Donald and Hamblin (1976) argued that no research was complete if measurements of vegetative mass did not accompany all measurements of yield. Estimating vegetative mass of crops that shed their leaves by maturity, such as soybean, is difficult, although Schapaugh and Wilcox (1980) demonstrated that variation of HI among soybean cultivars based on traditional measurements of the standing crop at maturity was highly correlated with harvest index estimates that included abscised leaves and petioles. The simplicity of the concept of HI, the amount of the total biomass partitioned to yield, and the implication that changes in partitioning would increase yield, suggested that it may be useful as a selection index for plant breeders (Donald and Hamblin, 1976).

Additional interest in HI was simulated by observations in several crops (wheat, barley, rice) that higher yields of improved cultivars were primarily a result of improved partitioning, with little or no increase in total biomass, i.e. the increase in yield was associated with a decrease in vegetative mass (see Evans, 1993, pp. 238–260 and Hay, 1995 for summaries). These findings were shocking, because they suggested that there had been little improvement in the inherent productivity of modern cultivars (i.e. total biomass was constant), with all of the substantial increase in yield over time coming from changes in partitioning, not increases in dry matter production. Suggestions (Austin, 1994) that there may be a maximum HI that cannot be exceeded raised questions about the future of yield improvement. It should be noted, however, that yield increased in some crops without changes in HI. Maize, for example, has shown essentially no change in HI, while yields increased substantially (Tollenaar, 1989; Duvick, 2005). In more recent years, total biomass has increased in wheat (Shearman *et al.*, 2005) and rice (Peng *et al.*, 2000), suggesting that fundamental changes in crop productivity, not just simple changes in partitioning, can drive yield improvement.

Harvest indices of crops growing under optimal conditions are usually close to 0.50 (Azam-Ali and Squire, 2002, p. 36). Stress greatly reduced harvest index (Azam Ali and Squire, 2002, p. 36), indicating that stress usually reduced seed yield more than vegetative mass, but stress could also increase HI if it reduced vegetative mass more than seed yield. There is substantial evidence that HI was much lower in most crops prior to domestication (Inoue and Tanaka, 1978, potato (*Solanum tuberosum*); Austin *et al.*, 1993, wheat; Fageria, 2007, rice).

Although HI has been used extensively to analyse the yield production process, especially with small grains, the concept has been criticized for conceptual problems and for not providing useful information about the mechanisms responsible for yield changes associated with increases in HI. The ratio contains seed yield in both the numerator and the denominator (equation 5.1), which can lead to spurious correlations between HI and yield if the variation in total biomass is much less than the variation in seed yield (Donald and Hamblin, 1976; Charles-Edwards, 1982; Klinkhamer *et al.*, 1992). Harvest index is influenced by variation in both of the components of the ratio (vegetative mass and yield), which further complicates interpretation of changes in the ratio. These concerns don't negate the fundamental concept of HI as a representation of the balance between vegetative and reproductive mass, which can vary independently because they are produced at different times during the crop's life cycle.

Harvest index is used as a measure of partitioning between vegetative and reproductive growth, but it provides no information about how the change in partitioning occurred (i.e. how assimilate was diverted from one sink to another). Thinking of an increase in HI as a 'transfer' or a direct partitioning of dry matter between vegetative and reproductive plant parts may be misleading and wrong, since vegetative growth and the production of yield during seed filling are separated in time. A direct partition could easily occur if the growth phases overlap and there is truly an option for assimilate to be translocated to vegetative or reproductive plant parts; overlap between vegetative growth and seed filling is possible, but it is very much the exception to the rule. It is impossible to know whether changes in the ratio represent a true change in partitioning (instead of assimilate going to vegetative growth it went to reproductive growth), or whether it is only an apparent change in partitioning resulting from unrelated changes in vegetative or seed mass.

Harvest index is similar to yield, in that it describes the final product but tells us nothing about how the final product level was achieved. In view of the problems of interpretation and lack of insight into changes in basic plant processes, Charles-Edwards (1982, pp. 111–112) suggested that 'it seems more logical, and the problems of improving grain yields more tractable, to look directly at the phenological, physiological, and environmental determinants of grain yield'. Following this logic, perhaps we can use our analysis of the yield production process to investigate changes in plant growth and development that might lead to changes in HI.

What options are there to explain the increases in harvest index associated with higher yield? Increasing partitioning to reproductive growth during the critical period for seed number determination (Murata's (1969) Stage Two) could increase seed number, as discussed previously, but, increasing the size of the yield container without increasing the wherewithal to fill it would be of little value. An obvious exception is a crop that is sink-limited (yield container is too small); yield would benefit from an increase in assimilate partitioning to reproductive growth during the critical period. A true change in partitioning between vegetative and reproductive growth during seed filling, which would increase yield and HI, is unlikely because seeds are probably already the primary sink during seed filling in modern cultivars. Vegetative growth continued during seed filling in some old cultivars of soybean (Gay *et al.*, 1980) and groundnut (Duncan *et al.*, 1978), but not in modern cultivars. Stopping vegetative growth before seed filling (i.e. changing partitioning) probably contributed to the higher yields of improved cultivars, but opportunities for continued improvement are probably limited.

The direct transfer of assimilate between vegetative and reproductive growth separated in time can occur if the plant accumulates storage materials (carbohydrates and N) during vegetative growth and utilizes them during seed filling. Such accumulation and utilization of storage materials occurs in most grain crops, although the contribution to yield is usually small (20–30%, as discussed earlier), suggesting that the potential for increasing yield by increasing the transfer of storage reserves is not great.

There are other mechanisms that could be responsible for increases in HI, but they represent only apparent changes in partitioning, not a real transfer of assimilate from vegetative to reproductive growth. It is obvious from equation 5.1 that any variation in yield that is not associated with a similar change in vegetative mass will cause a change in HI. Reducing the vegetative mass without changing yield will increase the harvest index. The maximum vegetative mass of most grain crops is directly related to the length of the vegetative growth phase, so early maturing cultivars will often have a smaller vegetative mass and, if yield does not change or changes by a smaller proportion, a larger HI. Conversely, a late-maturing cultivar with a long vegetative growth period will probably have a lower harvest index, since yield will probably not increase proportionately to vegetative mass.

The apparent harvest index ((yield)(yield + maximum vegetative mass)$^{-1}$) of soybean decreased as the length of the vegetative growth period (planting to growth stage R5) increased from about 65–100 days (Fig. 5.1). Longer vegetative growth periods produced larger maximum vegetative masses, which, when coupled with no corresponding increase in yield, caused a decline in apparent HI. Longer total growth durations (probably reflecting a longer vegetative growth period, see Egli, 2011) resulted in declining HIs in soybean, rice, sunflower (Fig. 5.2) and barley (Donald and Hamblin, 1976). Cultivar improvement of wheat in several countries resulted in earlier anthesis dates (Slafer *et al.*, 1994a) which could have contributed to the increase in HI by decreasing vegetative mass. Fischer and Palmer (1984) selected for shorter plants in maize, shortening the time to flowering (presumably decreasing vegetative mass) and increased HI, but yield also increased which contributed to the higher HI.

In some of these examples, the increase in HI was not necessarily associated with a change in yield, but only with a decrease in vegetative mass as a result of a shorter vegetative growth period. This variation in HI does not represent a real change in partitioning (i.e. assimilate going to seeds instead of vegetative plant parts); it is only an apparent change in partitioning representing dissimilar variation in yield and vegetative mass due to changes in the length of the vegetative growth period. Simply shortening the vegetative growth period did not provide a mechanism to 'transfer' dry matter from one stage to another; there was just less vegetative mass produced. This negative relationship between vegetative mass and HI is fostered by the disconnect between vegetative mass and yield (Fig. 5.1). Canopy photosynthesis reaches a maximum when the crop intercepts 95% or more of the incident solar radiation and, at that point, additional increases in leaf area or vegetative plant size will not increase canopy photosynthesis; larger plants will not necessarily produce more photosynthesis or yield. Conversely, as long as the crop achieves complete ground cover during reproductive growth, vegetative mass can decrease without reducing yield and HI will increase. This

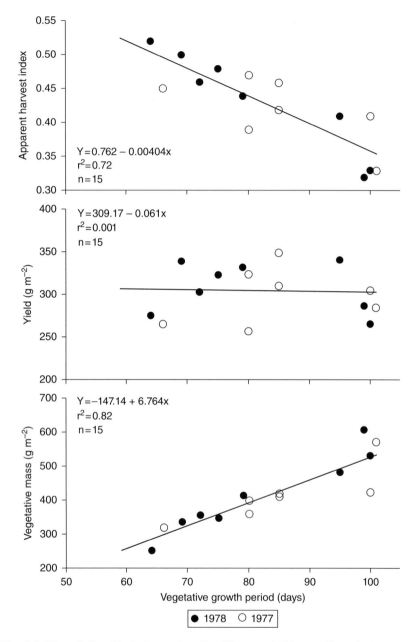

Fig. 5.1. The relationship between length of the vegetative growth period (planting to growth stage R5) and maximum vegetative mass, yield and apparent harvest index. Adapted from Zeiher *et al*. (1982). Eight soybean cultivars from maturity groups II to V were grown in the field for two years. Maximum vegetative mass was determined at the beginning of seed fill (growth stage R5). The apparent harvest index is the ratio of yield to maximum vegetative mass + yield.

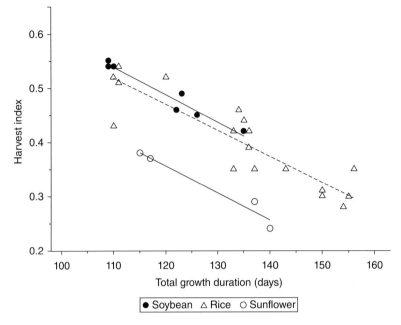

Fig. 5.2. Relationship between total growth duration (days from planting or emergence to maturity) and harvest index, for soybean (Schweitzer and Harper, 1985), rice (Venkateswarlu *et al.*, 1977) and sunflower (Villalobos *et al.*, 1994). Regression models: soybean: n = 8, Y = 1.095 − 0.005X, r^2 = 0.95***; rice, n = 21, Y = 1.049 − 0.0048X, r^2 = 0.83***; sunflower, n = 4, Y = 0.955 − 0.0050X, r^2 = 0.95*.

disconnect between vegetative or total growth duration and yield was discussed by Evans (1993, pp. 116–120) and Egli (2011) and is no doubt a characteristic of most grain crops.

Increasing yield with no change in vegetative mass via a longer seed-filling period would also increase HI (equation 5.1). As discussed previously, longer seed-filling periods are frequently associated with higher yields in many crops. Increases in SFD with no change in the length of the vegetative growth phase will result in later maturity, which could be undesirable in some environments. Earlier flowering and a shorter vegetative growth period (probably a smaller vegetative mass), however, could maintain the same maturity with a longer seed-filling period (Egli, 2004). Neither of these scenarios would necessarily result in an increase in vegetative mass, so any increase in yield would result in a larger HI. Fakorede and Mock (1978) (maize) and Sharma (1994) (wheat) reported increases in HI as SFD increased. There is ample documentation in the literature that improved cultivars in several crops owe their higher yields to longer seed-filling periods and independently, that modern cultivars frequently have higher HIs than older cultivars. Thus, it seems reasonable to speculate that some of the historical increases in harvest index came from longer seed-filling periods.

This mechanism provides an explanation for the constant biomass, higher yield and higher HI (Evans, 1993, pp. 238–260; Hay, 1995) scenario that stirred so much interest in HI. If vegetative mass decreased (shorter vegetative growth phase) in direct proportion to the increase in yield (longer seed-filling period), the total biomass (vegetative mass + seed yield) would stay the same, but HI would go up. Once again, this would not be a result of true changes in partitioning; instead, it would represent only an apparent change resulting from variation in the length of the vegetative and reproductive periods.

Environmental effects on HI (Donald and Hamblin, 1976) are to be expected, given the separation in time of vegetative and reproductive growth that potentially exposes then to different environments. Yield is not always closely associated with vegetative mass consequently environmental modifications may affect vegetative growth but not yield, or vice versa, causing changes in HI. Stress during seed filling, for example, would reduce yield without affecting vegetative mass, thereby decreasing the HI. An example of this differential response was seen when N fertilizer increased vegetative mass of wheat without affecting yield, causing a decrease in HI (Donald and Hamblin, 1976). Again, these changes in HI do not represent a true change in partitioning.

Although HI is commonly used to quantify changes in partitioning between vegetative and reproductive growth, it provides little information about the modifications of the plant that were responsible for the change. Focusing on the ratio distracts us from considering the plant growth processes directly involved in the production of yield and can lead to incorrect conclusions. In spite of the encouragement of Donald and Hamblin (1976), HI has not been widely used by plant breeders to develop higher yielding cultivars. Perhaps Charles-Edwards (1982, pp. 111–112) was correct: it is more beneficial to deal directly with the processes involved and ignore this simple ratio.

Time and Yield

Yield is always a function of a rate of biomass or seed dry matter accumulation expressed over a finite time. The rate of growth usually receives more attention from those interested in yield than time, although time is an equally important component of the yield production process. The association between the SFD and yield discussed earlier is an example of the importance of time. The direct association of time with yield is not the only aspect of time that is important, time also affects crop management strategies and the efficient utilization of available environmental resources. We cannot ignore time in our consideration of the yield production process.

Potential productivity

The potential productivity of any environment is determined by the solar radiation available when temperatures are suitable for crop growth (de Wit, 1967).

Grain crops cannot grow and produce yield without solar radiation as an energy source, and they cannot utilize this radiant energy unless temperatures fall within a suitable range for a period long enough to accommodate the life cycle of the crop. Solar radiation and time (period when temperatures are suitable for plant growth) set the upper limit of productivity at any location. These fundamental elements of the plant's environment are unalterable by man in a practical sense, in comparison to those elements that are frequently manipulated to increase crop productivity. Farmers routinely fertilize, irrigate, and control weeds, insects and diseases to increase crop yield, but the ultimate limit to yield will be determined by the solar energy available to drive photosynthesis when temperatures are suitable. In tropical climates with distinct wet and dry seasons, the availability of water may practically replace temperature in determining when crops can be grown (Goldsworthy, 1984). Our definition of potential productivity, however, excludes limitations from water deficiencies because irrigation could, and often does, eliminate the restrictions of the dry season.

Potential productivity should not be confused with potential yield or yield potential. Potential yield is defined as 'the yield of a cultivar when grown in environments to which it is adapted; with nutrients and water not limiting; and with pests, diseases, weeds, lodging and other stresses effectively controlled' (Evans, 1993, p. 212). Potential yield is a functional concept that describes the capacity of the plant to accumulate dry matter and produce yield in a particular environment in the complete absence of stress. Potential productivity essentially sets the energy available to the plant and the time it has to convert that energy into yield, while potential yield is a measure of how well the plant exploits the potential productivity in the absence of stress. This exploitation is controlled by the plant and temperature; all other limiting factors are removed by definition. A crop will only utilize a fraction of the potential productivity, even if it achieves its potential yield. The value of the potential productivity concept lies in comparison of locations and climates and, in my opinion, its focus on time as an important aspect of agricultural productivity.

Potential productivity is estimated by summing the daily solar radiation over the period when temperatures are suitable for plant growth, often taken as the length of the growing season (i.e. days from the last occurrence of $\leq 0°C$ in the spring and the first occurrence in the autumn). The time component is a major determinant of potential productivity, so potential productivity is largest in the tropics where crops can grow for 365 days in a year. It gradually declines with distance north and south of the tropics and eventually reaches zero for grain crops, when the growing season is too short for them to successfully complete their life cycle. Mean potential productivity across the middle of the US increased more than twofold, from approximately 2200 MJ m^{-2} at 49°N latitude (International Falls, MN) to 5000 MJ m^{-2} at 30°N (New Orleans, LA) (Fig. 5.3).

By comparison, mean potential productivity in the Cerrado region of Brazil (14°S) with a 365-day growing season was 6900 MJ m^{-2}, more than twice that in the heart of the highly productive US corn belt (40–45°N). Most of this variation is due to the time component with the average length of the growing season in the

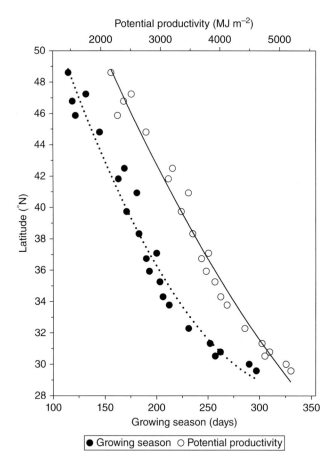

Fig. 5.3. Variation in time and potential productivity from north to south across the corn–soybean belt in central USA (~ 90°W longitude). The growing season is the average (1971–2000) number of days from the last frost (0° C) in the spring until the first frost in the autumn (NOAA, 2016). Potential productivity is the summation of average (1989–2008) daily solar radiation (MJ m⁻²) during the growing season (CRA, 2016).

US increasing by nearly 180 days (115–295 days) from north to south (Fig. 5.3) with another 70-day increase to the Brazilian Cerrado. Clearly, time is an important component of potential productivity.

The time available for crop production and potential productivity is also sensitive to elevation, decreasing as elevation increases with no change in latitude. The longer growing seasons associated with rising temperatures from climate change will also increase potential productivity and areas with enough time for grain crop production could expand to higher latitudes (Easterling, 2002; Chen *et al.*, 2012).

These data document the large differences in the time available for crop growth, but the question is: how well do our crops and agricultural production

systems exploit these differences in time and potential productivity? This exploitation is a function of the characteristics of the crop plant, the management system
and economic reality. We cannot consider this complex system in detail; instead,
we will focus on its basic component – the crop plant.

Utilization of potential productivity

The time used by an individual crop to produce yield varies significantly among
and within species. A summary of 13 grain crops from the literature found that
the total growth duration varied from only 62 (cowpea) to 185 days (sorghum)
(Fig. 5.4). There are reports of even longer durations, such as a sorghum land race
in Ethiopia that required 240 days to reach maturity (Mulatu and Belete, 2001),
rice cultivars that matured in 260 days (Grist, 1986, p. 90) and common bean cultivars that took 200 days to reach maturity (Graham and Ranalli, 1997). Whether
the species variation in Fig. 5.4 is related to characteristics of the species, their
area of origin, or is simply an artifact of the relatively small sample size for some
species is not clear. Regardless of its origin, this variation among species describes
a three-fold difference in potential resource capture.

Logically, one might expect longer growth durations to result in greater resource capture and higher yield. This supposition is correct if yield is defined as
total biomass at maturity; Murata (1981) found that the biomass of both C_3 and
C_4 species increased as the length of the growing season increased from 100 to 365
days (see Fig. 2 in Murata, 1981). The productivity of the C_4 species in Murata
(1981) was over 60 t dry matter ha^{-1} with a growth duration of 365 days, compared with 30 t dry matter ha^{-1} from a 125-day growth cycle. Similar relationships
between growth duration and total biomass were reported by Monteith (1978).
Crops that produce yield during much or all of their total growth cycle, (e.g. forage
plants, sugar cane (*Saccharum officinarum* L.), and tobacco) should benefit directly
from a longer growth cycle (Bunting, 1975).

Does the yield of grain crops follow this same pattern with crop productivity
and yield tracking the increase in potential productivity across latitude (Fig. 5.3)?
Average soybean and maize yields (2005–2014) from north to south across the
central US (Fig. 5.5) suggest that they do not. Maximum irrigated maize yields in
this transect occurred between 40 and 42°N (Nebraska) and then they decreased
at lower latitudes. The trend for non-irrigated soybean yield was similar to maize,
although the decrease at lower latitudes is not as large. Irrigation in Arkansas
(34–36°N) did not reverse the trend for lower yields at lower latitudes. The potential productivity and time available for crop growth increased at lower latitudes
(45% increase from 41 to 32°N) (Fig. 5.3), but yield did not increase (Fig. 5.5).
This trend was the same for irrigated and non-irrigated production, suggesting
that the failure of maize and soybean to utilize more of the potential productivity
at lower latitudes was not simply a function of water availability (shallow soils
with a greater likelihood of water stress). The lower yields at lower latitudes probably reflect the combined effect of poor-quality soils (low organic matter and low

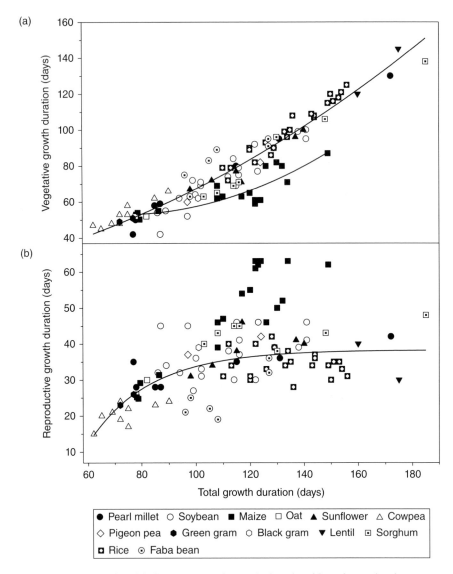

Fig. 5.4. The relationship between total growth duration (days from planting or emergence to maturity) and (a) the duration of vegetative growth (days from planting or emergence to anthesis or the beginning of seed filling) and (b) the duration of reproductive growth (anthesis or the beginning of seed filling to maturity) for 13 grain crop species. Adapted from Egli (2011). Original data sources can be found in Egli (2011). Regression models for (a); all species except maize, n = 86, Y = 18.68 + 021 X + 0.0027X2, R^2 = 0.94***; maize, n = 18, Y = 83.58 − 0.85X + 0.006X^2, R^2 = 0.71*** and for (b), all species except maize, n = 86, Y = −306.52 + 344.83 (1−e$^{(0.0432X)}$, R^2 = 0.43***.

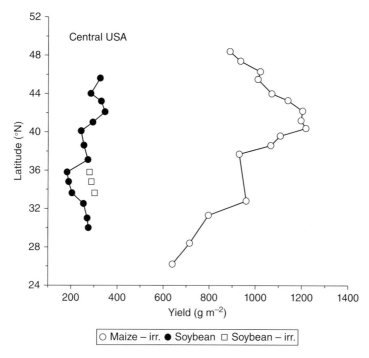

Fig. 5.5. Average maize and soybean yield (2005–2014) on a transect from north to south across Central USA. Soybean data are averages by select crop-reporting districts from Minnesota, Iowa, Missouri, Arkansas (irrigated data from Arkansas), and Louisiana. Irrigated maize data are from North Dakota, South Dakota, Nebraska, Oklahoma and North Texas. All data from NASS (2016).

fertility), higher temperatures and greater prevalence of diseases and insects, but even if the crops were grown in a perfect 'non-stress environment', it's unlikely that yields would be higher in the south than those in the north. The yield of these crops did not benefit from the greater potential productivity or the longer growing season at lower latitudes. They could not utilize the extra time and solar radiation to produce more yield.

Why can't soybean and maize, and, most likely, all other grain crops, produce higher yields in locations with long growing seasons and higher potential productivity? Cultivars with longer total growth periods are available (Fig. 5.4) (Egli, 2011) and were traditionally used in locations with long growing seasons, but apparently the long growth duration didn't contribute to higher yield. The basis for this failure can be found by considering how grain crops use time in the production of yield. Previously, we divided the total time required for growth of a grain crop into three stages, vegetative growth (Stage One), establishing the size of the yield container (Stage Two) and filling the yield container (Stage Three) (Murata, 1969).

The length of the vegetative growth phase (from planting to anthesis or the beginning of seed filling) was directly related to the total growth duration for 13 crop species, including both legumes and cereals (Fig. 5.4a). The reproductive duration (from the end of vegetative growth to maturity), however, reached a maximum at a total growth duration of 110–120 days (Fig 5.4b) and did not change as total growth duration increased to 185 days. A single regression equation described the responses of 12 of the 13 species; maize was the exception with a shorter vegetative phase and a longer reproductive phase than the other species. These relationships between vegetative and reproductive growth are similar to those described previously by Lawn and Imrie (1991) and Evans (1993, pp. 116–120).

The relationships in Fig. 5.4, based on data from the literature, were confirmed in several experiments featuring comparisons of cultivars with a range in maturity. Most of the variation in total growth duration (85–145 days) in an experiment with soybean cultivars from MG 00 to MG V was accounted for by variation in the length of the vegetative phase (Fig. 5.6). Seed-fill duration, estimated by the effective filling period (EFP), increased until the total growth duration reached 105 days and then remained constant as the total growth duration continued to increase. The longer total growth period resulted in a long vegetative growth period and, presumably, a larger vegetative mass, but a relatively constant SFD.

Soybean cultivars from maturity group (MG) 00 to VI (total growth duration 65 to 130 days) produced maximum yield when the total growth duration reached 100 days (Egli, 2011, original data from Edwards and Purcell, 2005a) as predicted by Fig. 5.4. Ishibashi et al. (2003) also reported increasing soybean yield as total growth duration increased from 77 to 90 days. In an irrigated field experiment, yield did not change as the vegetative mass increased from 372 (MG I) to 622 g m^{-2} (MG V) (Table 5.2). The maturity group V cultivar utilized the extra 40 days of growth to produce nearly twice as much vegetative mass as the MG I cultivar, but the larger vegetative mass made no contribution to yield. All of these data are consistent with the relationships shown in Fig. 5.4; short-duration cultivars used time much more efficiently by producing the same yield in less time than long-duration cultivars. Cultivars with longer total durations can be grown in areas with longer growing seasons, which may increase total resource capture, but they will not necessarily translate that extra time and resource capture into higher yield. Grain crops are relatively inefficient at using time to produce yield.

The disconnect between total growth duration and yield can be explained by the failure of the duration of canopy photosynthesis and reproductive growth to increase in step with the total duration. The larger vegetative mass associated with longer durations does not necessarily increase solar radiation interception so canopy photosynthesis would not increase. A constant canopy photosynthesis during reproductive growth with no change in the duration of reproductive growth provides little opportunity for yield to increase in concert with total duration. Reproductive growth reached its maximum length at durations of 110–115 days; longer total growth durations would increase vegetative mass, but there would be no change in yield.

The relatively short seed-filling period, at best only 30–40 days long, limits yield potential and contributes to the inefficient use of time by grain crops. At

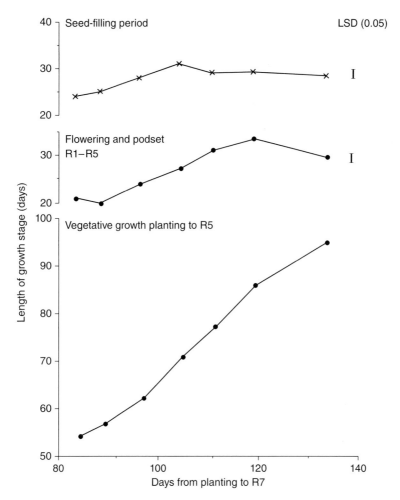

Fig. 5.6. The relationship between the length of vegetative and reproductive growth periods and the total growth cycle (planting to physiology maturity, growth stage R7) for soybean cultivars from maturity group 00 to V, averaged across four cultivars per maturity group and two years. All cultivars were grown in field at Lexington, Kentucky (38°N latitude) and the duration of seed filling was estimated by the effective filling period (EFP). Adapted from Egli (1994a).

the beginning of seed filling (Stage Three), no yield has been produced; vegetative growth and establishing the yield container are simply preliminary events. Producing high yields in such a short time places a heavy demand on the activity of the source during seed filling.

Root crops (e.g., potato or cassava (*Manihot esculenta* Crantz)) have a significant advantage over grain crops because they have much longer tuber- or root-growth durations (some in excess of 100 days (Wilson, 1977; Alves, 2002)).

Table 5.2. Effect of cultivar maturity on soybean yield. Adapted from Egli (1993).

Cultivar	Maturity group	Total growth cycle[1,2] (days)	Seed-fill duration[3] (days)		Yield[2] (g m^{-2})	Maximum vegetative mass[2,4] (g m^{-2})
			1989	1990		
McCall	00	82	24	25	274	305
Hardin	I	96	30	27	338	372
Harper	III	110	33	24	337	428
Essex	V	136	34	37	330	622

[1]Total growth cycle is from planting to physiological maturity (growth state R7).
[2]Average of 1989 and 1990, unshaded control plots. All plots were irrigated to minimize water stress.
[3]Effective filling period (final seed size/seed growth rate), LSD (0.05) for cultivar. comparisons was 7 days in 1989 and 4 days in 1990
[4]Total above-ground vegetative mass at growth stage R5, beginning seed fill.

The crop with the longest time to produce economic yield will usually have an advantage, unless the differences in time are counterbalanced by differences in growth rate due to environmental effects, an unlikely prospect when time differences are large. Record yields of potato (dry matter basis) are 73% higher than record wheat yields and 96% higher than record rice yields (Evans, 1993, p. 288). This advantage is probably partially related to the differences in seed (tuber) filling periods and it raises the interesting question of whether or not root and tuber crops should play a more prominent role in providing food for future populations.

Perennial grain crops are being developed to exploit more of the available growing season than annuals crops, i.e. be more efficient (see Chapter 1) (Glover et al., 2010; Taylor et al., 2013). The increase in total resource capture will not benefit yield unless the seed-filling period of perennial types is longer than annual types. The data in Fig. 5.4 suggests that higher yields from perennial grain crops may be an unrealistic expectation.

The inefficiency of a single grain crop in a long growing season is commonly overcome by growing more than one crop per year. Multiple cropping, widely used in many environments, increases the time devoted to the production of yield by essentially creating multiple seed-filling periods in a year. Multiple cropping is used successfully in temperate (e.g. growing soybean after winter wheat, Caviglia et al. (2004); Heatherly and Elmore (2004)) and tropical (e.g. rice–wheat systems, Timsina and Connor (2001) and multiple rice crops in a single year (Yoshida, 1977)) environments. Researchers at the International Rice Research Institute obtained a total rice yield of 23.7 t ha^{-1}, averaged across three years, from four crops in one 12-month period in the 1970s (Yoshida, 1977). The highest yield of a single crop in the sequence was 8.5 t ha^{-1}. Double cropping soybean after a winter wheat crop is popular in Kentucky and other southern states in the USA and also results in increased production per hectare per year. Interestingly, warmer temperatures associated with climate change will lengthen the growing season and increase the land area suitable for more than one crop per year (Yang *et al.*, 2015).

Although total productivity of multiple cropping systems is high, some of these systems are not as attractive from an economic standpoint. Yield of individual crops in a multiple cropping system may be less than when grown singly, while production costs for each crop may not be reduced as much as yield, resulting in a reduction in profits.

Short-duration rice and grain legume cultivars were the key to successful multiple cropping in many production systems (rice, Timsina and Connor, 2001, Kropft *et al.*, 1995; mung bean, Malik *et al.*, 1989; cowpea, Fisher, 2014). As discussed previously, shortening the vegetative growth period would not necessarily reduce yield, making it possible to have the best of both worlds – short duration for multiple cropping without sacrificing yield potential. Ratooning (managing regrowth from a harvested crop for grain or fodder) of, for example, sorghum, pearl millet, or rice (Plucknett *et al.*, 1970) or utilizing a second flush of pods on cowpea (Hall, 1999) also provides a second filling period and increases the productive use of available resources when time permits. The early maturity of short-duration cultivars can also provide an important source of food before the main crops reach maturity in subsistence production systems (Ehlers and Hall, 1997).

Professor Bunting addressed the question: 'Is your vegetative phase really necessary?' and concluded 'Yes, but not as necessary as one might think' (Bunting, 1971). The vegetative phase is needed to produce organs for nutrient absorption and photosynthesis, but, for grain crops, more vegetative growth is not necessarily better. Many grain crops in modern production systems, like soybean (Table 5.2), probably produce more vegetative growth than needed for maximum yield.

Although the length of the vegetative growth phase and the maximum vegetative mass are not necessarily related to seed yield, there are specific situations where a long duration/large vegetative mass may be beneficial. A large vegetative mass could increase stored carbohydrates or N that are available for redistribution to the seed (Egli, 1997). Redistributed assimilate contributes to yield (Evans, 1993, pp. 254–258), but the significance of this source of assimilate varies among species and environments, and is most important with late-season stress (Foulkes *et al.*, 2009). The generally poor correlation between maximum vegetative mass and yield suggests that redistributed assimilates are not always important, supporting the contention that a long growth cycle may not be necessary.

Long vegetative growth periods and large plants may have been necessary to produce the leaf area needed for maximum solar radiation interception in traditional cultural systems with wide rows (Dofing, 1997; Andrade *et al.*, 2002). Wide rows (1 m) were needed in these systems to facilitate crop production with horses (horses had to fit between the rows) and mechanical weed control. The common recommendation to use full-season cultivars, i.e. cultivars that use most of the available growing season to produce yield, in many cropping systems may be rooted in these traditional management practices. In modern cropping systems, however, mechanical cultivation is not needed and crops can be grown in narrow rows, so maximum radiation interception, maximum canopy photosynthesis, and maximum seed number can be achieved with short vegetative growth periods (early-maturing cultivars) and lower maximum LAI (Andrade *et al.*, 2002; Edwards and Purcell, 2005b; Edwards *et al.*, 2005). Modern narrow-row cultural systems have essentially eliminated the need for long vegetative growth periods with a large vegetative mass for many crops.

Cultivars with long vegetative growth periods may be more tolerant of stress during vegetative growth because reductions in growth and LAI can occur without reducing solar radiation interception and compromising yield (Lawn, 1989; Jiang and Egli, 1995; Andrade *et al.*, 2002). Genotypes with long vegetative growth periods often produce root systems that can extract water from deeper in the soil profile (sorghum, Blum and Arkin, 1984; sunflower, Gimenez and Fereres, 1986; barley, Mitchell *et al.*, 1996; soybean, Dardanelli *et al.*, 1997; Blum, 2009). Long vegetative growth periods provide more time for the plant to accumulate nutrients from infertile soils for ultimate redistribution to the seed (Wada and Cruz, 1989), resulting in a higher-quality seed. Long vegetative growth periods should provide an advantage in dual-use production systems, where the stover is used for construction material, biofuel production or as fodder after the seeds are harvested (Mulatu and Belete, 2001).

Long or short vegetative growth periods may be useful to position reproductive growth in a more favourable environment to avoid stress (Curtis, 1968; Bunting, 1971; Ludlow and Muchow, 1990; Azam-Ali and Squire, 2002, p. 76; Blum, 2009). The very successful Early Soybean Production System (ESPS) in the Mid-South region of the US uses early-maturing cultivars planted early to avoid late-season water stress resulting in dramatic increases in yield (Bowers, 1995; Heatherly, 1999; Edwards *et al.*, 2003). Short vegetative growth periods may, as mentioned previously, increase susceptibility to short-term stress early in vegetative growth, by increasing the probability that stress will reduce the LAI below the critical level.

Short-duration cultivars may reduce the total water requirement for irrigation without sacrificing yield (Ishibashi *et al.*, 2003; Edwards *et al.*, 2003, 2005), a consideration that will be more important as the competition for scarce water resources increases in the future (Wallace, 2000). On the other hand, long vegetative growth periods may have a negative effect in water-limited environments by exhausting water supplies during vegetative growth and leaving little for reproductive growth (Fischer, 1979).

There are many situations where either long- or short-duration cultivars provide a yield advantage or a more efficient useful production system. These advantages are specific for species, environments and cultural systems, but fostering multiple cropping and positioning reproductive growth in the most favourable environment may be two of the most important, while reducing water use may become more important in the water-limited environments of the future. The potential value of long- or short-duration cultivars should not be allowed to obscure the basic principle that there is no inherent relationship between yield and total growth duration once it exceeds the minimum needed for maximum yield.

Summary

Potential productivity (solar radiation available when temperatures are suitable for plant growth) is the basic resource that sets the upper limit of agricultural productivity. The time available for crop growth, controlled primarily by temperature, is a key component of potential productivity, and it decreases from 365 days in the tropics until there is not enough time for grain crop production at high latitudes. Unfortunately, grain crops are inefficient users of time and cannot convert time into yield when the time available for crop growth exceeds roughly 100 days. This inefficiency is often overcome by growing several crops in a single year at low latitudes. Climate change and warming temperatures will increase the time for crop growth in many locations, but it may be difficult to use this extra time to produce more yield.

Summary

In this chapter, we have considered the classic components of the yield-production process – seed number, seed size, source–sink relationships, partitioning and harvest index, and time – and their relationship to yield. Involving the developing seed in these considerations provides a deeper understanding of the role these processes play in the determination of yield, and it provides a framework to better evaluate these processes experimentally. Hopefully, this framework will make it easier to go beyond the 'apply a treatment and measure a response' approach to developing mechanistic understandings of the role these processes play in the production of yield.

Our analysis clearly illustrates that yield of all crops is limited by the source. Seed number is the primary determinant of yield and it is usually determined by the availability of assimilate from photosynthesis. Sink limitations can occur, but they are relatively infrequent and the magnitude of the limitation is hard to estimate. A source limitation means that increasing yield will require an increase in the rate and/or the duration of canopy photosynthesis. The latter option will require seeds that can grow longer. Although changes in partitioning contributed to higher yields when crops were first domesticated, our analysis suggests that the potential for future improvements in this area may be limited. When it comes to yield, the source rules!

The Way Forward

6

Yield Improvement

The world's food supply has always depended upon the productivity of grain crops – the land area harvested and the yield per unit area. Converting the production of these grains into food on the tables of the world is a complex and convoluted process, with many social, cultural and economic ramifications, some of which we will discuss later in this chapter. Improvements in agricultural productivity have kept the world reasonably well fed for at least the last half-century or so, as the world population increased from 3 to 7.3 billion (FAOSTAT, 2016). Adequate food supplies are still an important issue in the early years of the 21st century, as we face the challenge of feeding a population that will probably approach 10 billion by 2050, just 34 years from now. This increase in food demand (both quantity and quality) must be met against a background of a changing climate, declining availability of natural resources, including water, and a population that will be increasingly concentrated in urban areas. The productivity of grain crops will have a central role in this drama, as they have had since the beginning of agriculture some 10,000 years ago. There is general agreement that future increases in productivity will come mostly from higher yields, since the opportunities to increase the land area devoted to crop production are limited. The focus of this chapter is on yield improvement and the question of what is the best way forward to einsure a well-fed world in 2050.

The dramatic increase in yield of most grain crops since the middle of the last century (see Chapter 1) was primarily driven by genetic improvement of the plant and changes in crop management practices (improvements in the plant's environment). Smaller contributions may have come from changes in where crops were grown (shift from low-yielding to high-yielding environments) (Beddow and Pardley, 2015), what crop species were utilized, and increases in the CO_2 concentration in the atmosphere (Hatfield *et al.*, 2011). The relative contribution of breeding vs management is debated in the literature, with estimates ranging from 20% to 80% of the increase coming from plant breeding (estimates for the major crops in the US are usually close to 50%; see Chapter 1). Regardless of the debate, all would agree that both approaches contributed, with each being relatively in-effective by itself, as shown by numerous examples of cultivars that express their high yield potential only when appropriate management practices are deployed. It has been suggested, however, that opportunities for future increases from man-agement may be limited in highly developed agricultural systems (Egli, 2008; Fischer *et al.*, 2014, p. 358).

Improvements from management are limited because each improvement in the plant's environment makes future increases more difficult (e.g. when weeds are controlled, fertility deficiencies eliminated or row spacing and plant population op-timized, no further improvement is possible). Management, however, will continue to contribute to improvements in the efficiency of production, an important aspect of any production system, but one that does not necessarily contribute to higher yield. Precision agriculture approaches provide many new opportunities to increase efficiency by, for example, adjusting inputs to match soil productivity levels, thereby reducing input costs without reducing yield (Yang *et al.*, 2016). Improvements in efficiency will also contribute to reducing the environmental impact of agriculture. It seems to be more difficult to use these high-tech precision approaches to increase yield; perhaps more yield enhancements will come as the technology matures. It is too early to evaluate the contribution of the precision management practices asso-ciated with the 'big data' approaches currently offered by agribusiness.

Failure to deploy appropriate management technologies that are available also limits yield. This failure is always present to some degree (every producer cannot always apply the best management practices at the appropriate time), given the time sensitivity of many management practices, the complications asso-ciated with managing large areas, and the vagaries of the weather. Development of management practices that simplify or reduce the need for timely deployment could increase productivity by 'better' application of existing technology. For ex-ample, Roundup Ready technology probably improved weed control and yield by simplifying the weed-control process, even though excellent weed control was possible before the technology was introduced. Economic and cultural limitations often prevent use of appropriate management technology in less-developed agri-cultural systems, but its application in these systems can result in dramatic in-creases in yield. Failure to utilize available technology is a separate issue from the development of new management practices that will increase yield. Securing the application of available technology, even in the face of economic restraints, is usu-ally simpler than developing new management technologies.

Historically, plant breeders improved crop plants by selecting for higher po-tential yield and by eliminating plant defects that they thought reduced yield. Defect elimination did not increase potential yield, it simply restored yield to the level that would have occurred in the absence of the defect. Disease resistance, for example, simply restored yield to the level that occurred in the absence of the dis-ease. A dramatic example of defect elimination occurred when the yield increase in a comparison of barley cultivars from different eras was greatly reduced when leaf diseases were controlled with fungicides (Sandfaer and Haahr, 1975). Defect elimination becomes progressively more difficult with each cycle of improvement.

Shattering and lodging were early targets in many crops, but, once these defects were eliminated, the next improvements were more difficult and they will stop, in theory, when the 'perfect' plant is produced. Changes in the environment, and disease and insect pressures, will probably preclude the production of this 'perfect' plant and insure a constant need for defect elimination in the future.

Both approaches increase yield in the farmer's fields, but it is difficult (and probably of interest only to plant breeders and crop physiologists) to estimate the contribution of the two approaches, which certainly varies by crop, location and era. Some have argued that all of the yield increases in maize were due to defect elimination (assuming that improving stress tolerance is a form of defect elimination), with no change in potential yield (Duvick and Cassman, 1999; Tollenaar and Wu, 1999), but now there is substantial evidence for contributions from improvements in yield potential (Egli and Hatfield, 2014a,b; Fischer *et al.*, 2014, pp. 542–547). The longer seed-fill duration that resulted from selection for yield in many crops, including maize, is an example of improvement in yield potential (see discussion in Chapter 5).

Plant breeders successfully increased yield potential by selecting for yield without considering the physiological processes or plant characteristics involved in the production of yield. Defect elimination involves selecting for very specific plant traits, but defect elimination, by definition, does not increase yield potential. In a landmark paper in 1968, Donald (1968) suggested that breeding progress could be increased by defining the plant characteristics that contribute to higher yield (i.e. developing an ideotype of a high-yielding cultivar) and then selecting for those characteristics, instead of blindly selecting for the final product – yield.

Donald's (1968) initial wheat ideotype was primarily based on whole-plant characteristics (leaf angle, leaf size, tillering etc.), but it was easy to translate his ideotype approach to physiological processes involved in the production of yield. Actually, plant breeders have always used a form of ideotype breeding when they selected for traits (besides yield) that they felt were needed in high-yielding genotypes (e.g. lodging, shattering and disease resistance), but these traits were mostly related to defect elimination and, at best, were only a loose, informal utilization of the ideotype approach. Relatively modest heritabilities of yield and substantial genotype by environment interactions have always limited direct selection for yield. Selection for specific traits, which could reduce the genotype by environment interaction, leading to more rapid progress, was thought to be an improvement over yield-based approaches. The potential opportunity for greater improvements in yield and the presumption that the ideotype approach was more scientific, and therefore better than simply selecting for yield, encouraged crop and plant physiologists to identify traits that could be used to increase yield (e.g. Araus *et al.*, 2002). Crop physiologists used ideotype logic to justify their efforts to better understand the yield production process at the enzyme, process, whole-plant and community levels, and they were successful. The developments in molecular biology and our ability to manipulate individual genes also created a potential demand for a fundamental understanding of the yield production process.

Our understanding of the yield production process today is much more detailed at several levels than it was when plant breeders started actively improving crop yields early in the 20th century. Has the knowledge produced by crop physiologists contributed to historical increases in yield or is crop physiology only a retrospective science, as suggested by Evans (1975) and Miflin (2000), simply explaining the basis for past yield increases? The discipline of crop physiology is useful only to the extent that it is predictive, i.e. the knowledge and understanding it generates leads to higher yields via plant breeding or crop management, or more efficient production systems. The rapid growth of yield that started in the middle of the last century began when our understanding of the yield production process was very rudimentary compared to current levels, suggesting that yield improvement is not absolutely dependent on understanding the process. Evans (1993, p. 266) concluded that 'selection for greater yield potential has not, could not and never shall wait for our fuller understanding of its functional basis, despite pleas of physiologists...'.

The literature is full of reports from crop physiologists describing traits that they believed could be used by plant breeders to improve yield. A significant association between a specific process or trait and yield in an experiment was often all that was needed to suggest that the trait would be useful. It is much harder to document the use of such a trait to develop high-yielding cultivars that are commercially successful (i.e. yield more than the best cultivars available to producers). Sinclair *et al.* (2004) and Richards (2006) described several examples of successful utilization of specific traits in cultivar improvement programmes, including tolerance to high temperatures in cowpea, drought tolerance of N_2 fixation in soybean, improved water-use efficiency in wheat and shortening the anthesis–silking interval in maize under drought stress. Interestingly, all of these examples involved stress relief in some form, certainly important in yield improvement, but not, by definition, a part of potential yield.

Many of the supposedly useful traits defined by crop physiologists have been ignored by plant breeders and, when they were not ignored, attempts to use them to increase yield often failed. Direct selection for single-leaf photosynthesis has not been successful (Ford *et al.*, 1983), in spite of the fact that higher yields must be associated with increased resource capture, which requires, in the absence of having more time to accumulate dry matter, a higher rate of canopy photosynthesis. Genetic yield increases can often be traced to a longer seed-filling period (i.e. more time for resource capture) and plant breeders lengthened the seed-filling period by direct selection in several crops (see Chapter 5 and Egli, 2004 for a summary) and, inadvertently, when selecting for yield, but it was not useful in trait-based cultivar development (Pfeiffer *et al.*, 1991). Selection based on a new plant ideotype in rice, developed from physiological principles, did not increase yield (Peng and Khush, 2003). These examples, and there are many more that could be cited, illustrate the difficulty of using selection for specific traits to increase yield. Interestingly, in spite of support for the ideotype approach (Sedgley, 1991; Marte *et al.*, 2015), there is little evidence that it has been widely used in yield improvement programmes.

Why has directed trait-based plant breeding (ideotype approach) not made larger contributions to yield improvement? At first glance, it would seem that a breeding programme based on selection for specific traits that determine yield would lead to more progress than just selecting for yield, but it hasn't necessarily worked that way. The large populations needed when selecting for multiple traits may have limited adoption of this approach (Marshall, 1991). The difficulties associated with finding simple traits that determine yield has limited progress and the approach cannot be successful if the traits used do not control productivity. Many trait–yield associations identified by crop physiologists were based on simple correlative relationships, often from experiments with only a few genotypes. Additional studies to demonstrate cause and effect in a field environment are more difficult and were seldom carried out; consequently, direct selection for the trait often failed because there was no true relationship to yield. Some traits purported to control yield were evaluated on isolated plants in greenhouses and the advantage bestowed by those traits disappeared when the plants were grown in a community in the field.

Some traits required complicated techniques to quantify them, which made it difficult to use them in plant breeding programmes to develop improved cultivars. Crop physiologists demonstrate trait–yield relationships and evaluate genetic variation in precise experiments, often in controlled environments. Plant breeders, however, must evaluate the trait on large numbers of plants in the field, where lack of precision may obscure variation in the trait, making selection ineffective and resulting in no change in yield. The effort required to measure complex traits on large numbers of plants often discourages plant breeders from using trait-based approaches. High-throughput phenotyping approaches that automate trait characterization (Andrade-Sanchez et al., 2014; Crain et al., 2016) make it possible to easily deal with large numbers of plants, but they are often limited by the few traits that can be measured.

Trait-based selection programmes were historically limited to working with the natural variation within a species, which was, for many physiological traits, often relatively small, limiting their practical usefulness. This limited variability in physiological traits related to productivity is perhaps to be expected, as evolution probably evaluated and discarded inferior versions of many of these traits (Denison, 2012, pp. 28–42). Seed growth rate and the associated variation in seed size, two traits discussed at length in this book, exhibit substantial genetic variation within most species, but they are not related to yield (see Chapter 4); apparently, evolution over the millennia did not diminish the variation.

Advances in molecular biology, making possible the inter- and intra-species transfer of single genes, were thought to herald a new era of plant improvement. Plant breeders would no longer be limited by the available intra-species variation in desirable traits; genes conditioning higher productivity in any organism could be theoretically assembled into a single genotype. These developments stimulated searches for genes that would to be useful to increase yield and reports of reputed successful searches appear regularly in the literature (see Van Camp (2005), Zhang, (2007) and Dunwell (2010) for examples), but I know of no grain crop cultivars engineered specifically to produce higher yield that are currently available to producers. The herbicide-tolerant and insect-resistant maize, soybean and canola cultivars widely grown by grain producers may result in higher yield, but only because they are associated with better weed control or less insect damage; they would not yield more in a perfect environment, i.e. yield potential has not increased. Ambitious attempts are underway to move an entire plant process, C_4 photosynthesis, into a C_3 species (rice); participants are optimistic (von Caemmerer *et al.*, 2012), but it is too soon to know if and when it will be successful. Thirty-plus years into the biotech era, it seems that direct genetic engineering of crop plants to increase yield potential is still in the future. The initial optimism of biotechnologists was not warranted.

Yield is the cumulative result of the activity of all of the processes necessary for plant growth and their interaction with the above- and below-ground environment. Selection for yield targets the end result of this process, whereby trait-based or gene-based approaches select for only a single component of the system. Perhaps it is unrealistic to think that a single gene could have a significant effect on yield (Fischer, 2011). The failure of ideotype or biotech approaches to date suggests that we have yet to learn how to deal with such a complex system at the single trait or gene level. Perhaps the greatest contribution of molecular biology to yield improvement will come through enhancements of conventional breeding techniques to make them more effective. For example, the use of genomic selection may increase the rate of gain in yield substantially over conventional breeding approaches (Bassi *et al.*, 2016).

Efficiency (yield improvement per unit time) is also an important consideration in the debate between selection for yield- and trait-based programmes. Trait-based programmes will be useful only if their cultivars are higher yielding than those produced by plant breeders selecting for yield in the same time frame. A trait-based programme that increases yield has accomplished little if the resulting cultivars still yield less than those from competing yield-based programmes. Since plant-breeding progress is partially a function of the number of crosses and the size of the populations evaluated, selection for yield may have an advantage over trait-based programmes requiring evaluation of more complex traits. For example, higher-yielding spring wheat cultivars selected for yield were successful because their roots penetrated deeper into the soil profile (Pask and Reynolds, 2013). Would selecting directly for deeper roots produce comparable cultivars in the same time frame? Given the complexities of measuring roots, it may not have. Any consideration of yield improvement systems must consider efficiency and time; increasing yield is not the only criterion, but how long it takes to achieve a given level of improvement is also important. One of the advantages touted for genomic selection is its ability to make selections earlier in the breeding cycle, thereby increasing the rate of gain (Bassi *et al.*, 2016). Improvements in planting and harvesting technology are making it easier to handle large breeding populations, which provide an additional advantage to conventional breeding approaches.

Sinclair *et al.* (2004) suggested that successful trait-based approaches to cultivar improvement involved multi-disciplinary efforts sustained over many years, which could represent a significant impediment to the use of this approach. Multi-disciplinary efforts are hard to organize and harder to sustain in public institutions, given the current funding climate, with short funding cycles and few funds available for public plant-breeding programmes. Private industry has the resources to create and sustain trait-based programmes for long periods, but whether or not they are using these approaches, or whether they have been successful, is not usually public knowledge.

At this juncture, selecting the best approach to ensure continued increases in yield could be the key to future food security. Traditional breeding programmes based on selection for yield have a strong track record, driving yields upward since at least the middle of the 20th century. On the other hand, the ability to transfer genes among and within species, and edit genomes, provides unprecedented opportunities to manipulate plants. Should successful traditional breeding approaches be abandoned in favour of new, unproven technologies that have great potential? The biotech world has a reputation for not delivering on its promises, and some observers have suggested that manipulating individual genes to increase yield is unworkable and reflects 'excessive naivety with respect to the complex physiology of yield determination' (Fischer, 2011). Even if the biotech approach is successful, current public dissatisfaction with GMO foods could limit deployment of high-yielding GMO cultivars. Denison (2012, p. 4) expressed concern that emphasizing single traits and genes in trait-based and biotechnological approaches to yield improvement may actually limit progress by diverting funds from more traditional approaches.

Fischer (2011) pointed out that the biggest threats to world food security will come in the next 20 years, assuming that population growth drops to negligible levels by mid-century. Already, for example, the area devoted to rice production in Japan is decreasing as the population decreases (Normile, 2016), reducing the need for continued increases in yield. If this is the future for other areas of the world, relying on long-term projects to increase yield (trait-based selection programmes or gene-to-phenotype approaches) may produce major impacts after the need for them has largely gone away, i.e. after population growth slows dramatically. Selection for yield has had an excellent track record for nearly a century, and it may be our best hope in the short run, especially if enhanced by molecular approaches (QTLs, molecular markers, genomic selection, etc.). It seems foolish to completely abandon the yield-based approach until the utility of trait-based approaches is clear.

Food Availability for the Future

Current estimates suggest that food production must increase by 60–100% to feed the world in 2050, just 34 years from the present (Fischer *et al.*, 2014, p. 14; Hatfield and Walthall, 2015). These estimates are based on a 33% increase in population (7.3 to 9.7 billion, median estimate, UN Population Division, 2015) and rising affluence of some segments of the world's population, which results in more meat in the diet, thereby requiring greater grain production to provide the same number of calories per person. The diversion of food crops to produce bio-fuels places an additional burden on the food production system.

It is usually assumed that yield will be the primary driver of these increases, since land utilization for crop production is already near maximum levels, and significant expansion may not be possible without negative consequences (e.g. destroying forests, ploughing up permanent pastures, utilizing low-quality soils in fragile ecosystems) (Hall and Richards, 2013). Given this restraint, can yield be in-creased enough to supply the production needed to meet future demands for food?

First, it is interesting to ask whether there are any historical precedents for a yield increase of 60–100% in a 34-year period. Increases in world maize, rice and wheat yields for the 34-year period from 1979 to 2013 were 62% to 76% (FAOSTAT, 2016). Increases in maize yield in highly developed, very productive agricultural systems ranged from 47% (Iowa, USA unirrigated) to 66% (Nebraska, USA irrigated) over the same period (NASS, 2016). The increase in rice yield in Asia during the Green Revolution (1961 to 1983) was 73% (FAOSTAT, 2016). These increases often exceed the 60% increase thought to be the minimum needed to maintain adequate food supplies (assuming no change in the land area devoted to grain crops) through to 2050; it is encouraging that 'business as usual' without dramatic increases in yield growth rate may suffice for the next 34 years.

Food supply and demand relationships are often evaluated by comparing relative growth rates of yield and population. This comparison does not account for changes in production area and it may be difficult to appreciate the potential effects of small differences in relative growth rates extended over time. The ratio of total yearly production of a crop to population (kg per capita) captures the direct relationship between production (yield × harvested area) and the size of the population, and provides a single index to easily characterize the sufficiency of production. This simple index, however, does not account for all facets of the supply for a particular area in a given year. Imports, exports, accumulation or utilization of storage reserves, non-food uses, and waste between production and consumption affect food availability, but they are not part of the index. The suffi-ciency of any particular level of the index can be quite variable, depending upon consumption levels, dietary habits, and food choices, including the proportion of animal-based foods in the diet. In spite of these deficiencies, the index is useful, in my opinion, to characterize changes in the relationship between production and population, especially if we focus on the trends and don't attach too much meaning to the absolute values of the index.

The ratio for rice production in Asia increased substantially from 116 kg/capita in 1961 to nearly 150 kg/capita by the mid-1980s (29% increase) as a result of the effects of the Green Revolution on yield and a 17% increase in harvested area Fig. 6.1). The upward trend stopped in 1985 and then the ratio fluctuated around a relatively constant value through 2005, indicating that, on average, the increase in production (as a result of higher yield and a 14% increase in harvested

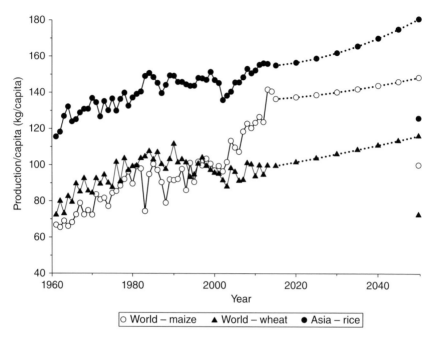

Fig. 6.1. Production per capita of wheat, maize and rice (solid lines) from 1961 to 2013). The projected production per capita (dotted lines) was calculated by assuming that the harvested area did not change from the 2012–2014 average and that yield growth continued at the linear rate estimated from 2005 to 2014. Total production (yield × harvested area) was divided by the median population estimates from the 2015 projections by the United Nations Population Group. Individual symbols at 2050 represent a no-growth scenario calculated from the average total production in 2012–2014 and 2050 population. All data from FAOSTAT (2016).

area) matched the increase in population. It is not yet clear whether the uptick after 2005 establishes a new high or whether it is just a result of year-to-year variation in the size of the crop. The ratios for world production of wheat and maize followed similar patterns (Fig. 6.1). The ratio for wheat declined from a maximum in the 1980s before apparently stabilizing at a somewhat lower level after 2000. The ratio for maize started to rise dramatically after 2002, reaching nearly 140 kg/ capita by 2013, partially as the result of a 48% increase in harvested area. The extensive use of maize for biofuel production (ethanol) was probably responsible for this increase in area; it is unlikely that this increase will continue. The ratio for rice is probably a better indication of food sufficiency than those of wheat and maize, since rice is primarily a food crop, with minimal use for feed or industrial products. The ratios for maize and wheat are meaningful, if we assume that there are no significant changes in the magnitude of non-food uses, but this assumption is not always valid, as shown by the increasing use of maize to produce ethanol that occurred in the early 2000s.

Per capita supplies of all three crops at the beginning of the 21st century were clearly much better than they were in 1961; more importantly, these trends suggest that historical yield and area increases of these crops roughly matched the 43% increase in population between 1985 and 2010. The pronouncements of doom that started with Malthus (1798) 216 years ago are still incorrect; agricultural production continues to this day to match growth in population.

These historical trends are encouraging, but what of the future? Will we finally approach the Malthusian apocalypse; will future production continue to increase in step with population or will there have to be revolutionary increases in yield to feed the world of the future? Predicting the future is always dangerous and most predictions end up being wrong. Predictions are based on assumptions, so it is easy to predict disaster or a rosy future by making the appropriate assumptions. In spite of this dismal view of the value of predictions, I will add mine to the many available. At the very least, predictions and the assumptions underlying them provide a useful framework to think about the future.

To estimate the capacity of the Asian rice crop to meet the needs of future populations, I calculated total production by assuming a continuation of the rate of growth of rice yield (54 kg ha^{-1} year^{-1} from 2005 to 2014) with no change in harvested area (average for 2012–2014 was 143 million ha) and then I used the median population estimates from the UN Population Group's 2015 report (FAOSTAT, 2016) to make the predictions shown in Fig 6.1. In comparison with 2013, production per capita increased by 14% by 2050, when the population in Asia is forecast to reach 5.3 billion (22% increase from 2013), suggesting that simply maintaining recent yield growth rates with no increase in harvested area will more than suffice for the most likely future increase in population. Applying the same approach to world production of wheat and maize resulted in a 16% increase in the ratio for wheat and a 6% increase for maize by 2050. Interestingly, assuming no growth in yield and a constant harvested area (constant total production) through to 2050 reduced production per capita for rice and wheat to levels that are substantially lower than those in 2013, but equal to (wheat) or still slightly higher (rice) than they were in 1961 (Fig. 6.1). The maize ratio was essentially equal to 2000 levels, reflecting the substantial increase in harvested area since 2000.

The assumptions underlying the predictions in Fig. 6.1 produced 40–60% increases in total production of the three crops by 2050; increases that are just below the often-hypothesized 60–100% increases needed by 2050 (Hatfield and Walthall, 2015). A 100% increase in total production by 2050 would produce ratios of 145, 200 and 252 kg/capita for world wheat, maize and Asia rice, respectively; ratios that are 32–43% higher than the projections in Fig. 6.1.

Obviously, estimates of the adequacy of future food supplies are directly dependent upon assumptions of future changes in production. Maintaining recent absolute yield growth rates through to 2050 slightly increased the ratios (Fig. 6.1), but will these modest increases suffice, or will the much larger ratios resulting from doubling production be required? The need for large increases in production is partially based on hypothesized increases in consumption, especially of animal products, but this is an 'optional' increase, it doesn't have to occur to maintain an adequately fed population, as defined by the ratios in 2010. In fact, the experience of developed countries suggests that increases in consumption will probably lead to poorer health and decreased well-being of the population. Analysis of temporal changes in food sufficiency, as expressed by these ratios, provides, in my opinion, an encouraging perspective on the prospects of feeding 9.7 billion people in 2050. Simply maintaining recent yield growth rates, with no change in harvested area, was more than adequate to maintain per capita food supplies through to 2050 with the most likely rate of population growth. Perhaps no revolutionary changes in the rate of yield improvement will be required in the future.

The extreme assumption of no growth in productivity resulted in ratios that were slightly higher than those in 1961 for wheat and rice, and roughly equal to the 1985 ratio for maize; wheat and rice ratios were no worse than what the world survived on in the middle of the last century. Surprisingly, the no-growth scenario did not result in catastrophic reductions in the food supply per capita.

Maintaining yield growth rates that have been relatively constant for at least the past 50-plus years seems to be more tractable than increasing them, particularly in the short term. It should be noted that maintaining a constant absolute growth rate (kg ha^{-1} year^{-1}) results in a steadily declining relative growth rate (% per annum), so the index suggests that the status quo can be maintained with a relative growth rate that is constantly decreasing, for example, from 1.2% per annum in 2011 to 0.8% in 2050 for rice in Asia. Hall and Richards (2013) and Fischer et al. (2014, p. 558) suggested that a rate of 1.1 to 1.3% per annum may be needed to meet future food demands, but Ray et al. (2013) concluded that a rate of 2.4% was needed to double production by 2050. Maintaining a constant relative growth requires a steadily increasing absolute growth rate, and there is no historical precedent for this in any crop. Constantly increasing the absolute growth rate will be a much bigger challenge to all aspects of the yield improvement process than just maintaining a constant absolute growth rate.

Evaluating the ratio of production to population (Fig. 6.1), however, does not consider unequal food distribution or that proportion of the population that is currently under-nourished. These two somewhat separate issues are important, but their resolution involves complicated social, cultural and economic issues. These issues are probably not as closely related to overall production levels as they are to production and distribution in specific countries, which is not reflected in the ratio in Fig. 6.1. In spite of these complications, the ratio of production to population provides a useful index of the capacity of production to meet the needs of the population. More importantly, the modest assumptions underlying the predictions in Fig. 6.1 produced an increasing ratio, not a decreasing ratio, suggesting that the problems of feeding future populations may not be as intractable as often described.

Interestingly, not everyone accepts the argument that large increases in food production are needed; some argue that the world is awash with food and the real issues are distribution and the wherewithal to buy food. If people had money they could buy food, so the solution to the food-supply problem lies in economic development (Bittman, 2014). Failure to always find a close association between total food production and food deficiencies supports this position. The world per capita food supply in 2011 was roughly 2800 calories/capita/day (FAOSTAT, 2016), surely an ample supply, but there were still many hungry people in the world. Many of the great historical famines were not necessarily related to production shortfalls; rather, they occurred when disruptions by war, natural disaster, high prices or political chaos reduced food availability. For example, the devastating famine in Bengal, India in 1943 that killed 2 million people occurred when the effect of a drop in rice production was greatly amplified by hyperinflation and high prices (O'Grady, 2009).

The near-impossibility of ever achieving a uniform global distribution of food to all peoples, coupled with the expected growth in food demand by 2050, makes it unlikely, in my opinion, that the world can be fed adequately without future increases in food production. As seen in Fig. 6.1, maintaining constant production levels would result in significant declines in production/capita by 2050. While sharing food equally among all peoples may be adequate today, it will probably not ensure a well fed population in 2050.

The problem of feeding the 9.7 billion is less daunting when we consider that increasing food production (yield and/or harvested area) is not the only option. While a large increase in land area used to produce crops is probably not a viable option, there may be some opportunity for further increases without cutting down tropical forests or expanding onto fragile soils. These opportunities are likely to be relatively small, perhaps in the order of 10% above 2008–2010 levels by 2050 (Fischer *et al.*, 2014, pp. 15–18). Less land will be required for food production in counties with decreasing populations, as is currently occurring in Japan (Normile, 2016). This 'extra' land could be used for production for export, although, in Japan this land is being abandoned.

The area producing crops can be effectively increased, however, by producing several crops on the same area in one year (Egli, 2011; Gregory and George, 2011; Andrade *et al.*, 2015), thereby increasing total productivity per year. Multiple-cropping systems are widely used (e.g. winter wheat–soybean in southeastern USA; two or three crops of rice in China; winter wheat–rice in Northern India, Pakistan, Nepal and Southern China (Cassman, 1999)) to effectively utilize the long growing season in tropical and semi-tropical areas. Double or triple cropping usually does not double or triple annual productivity, because the second and third crops may be grown in a less desirable environment (e.g. soybean grown after wheat usually yield less than single-crop soybean because they are planted after the optimum date (Egli and Cornelius, 2009)), but total productivity is higher than from a single crop. The length of the growing season limits the opportunities for multiple cropping, but it could be expanded beyond current levels in many locations. Interestingly, rising global temperatures and longer growing seasons (Hatfield *et al.*, 2011) should increase opportunities for multiple cropping.

The land area available for food production can also be effectively increased by abandoning use of crop land to produce biofuels. It is very unlikely, in my opinion, that a world with a population approaching 10 billion will be able to devote significant land area to the production of biofuels. The area needed to make a significant contribution to total energy use is so large that it would seriously reduce food production. Converting the entire US maize crop to the production of ethanol, for example, would only replace 18% of the country's annual gasoline use (Biello, 2011). The popular concept that significant quantities of biofuels can be produced on land unfit to produce food (Campbell *et al.*, 2008) is mostly fantasy; producing significant quantities of biofuels will certainly use land that could be used for food production, crop land that will eventually be required to feed the expanding world population.

Producing feed for the 145 million dogs and cats (ASPCA, 2016) and 10 million horses in the USA (FAOSTAT, 2016) also reduces land available to produce food, as does production of plants for industrial products (e.g. cotton or hemp for fibre). The opportunities for increasing the land area for food production by reducing competition from non-food uses will vary widely among production systems and countries, but it could make a small, but possibly significant contribution to increasing the food supply.

Farming the ocean provides another source of food that does not compete directly for scarce land supplies. Ocean-based fish-production systems are currently operating successfully, and no doubt could be expanded in the future (Bourne, 2015, pp. 166–181). The current opposition to consuming farm-grown fish would probably decrease as production systems improve and the price of competing sources of protein increase.

Increasing direct consumption of plants and decreasing meat in the diet would significantly increase food availability (Foley *et al.*, 2011). Present trends are in the opposite direction, driven by increasing affluence in some countries; these trends are responsible for estimates of future increases in food production that are higher than justified by population growth. Beef consumption per capita varies widely among countries, suggesting that there is ample opportunity for reductions in countries with high levels of consumption. Insects are an excellent source of protein with high ratios of protein produced to feed consumed, and are consumed in some societies. Increased use of insects could replace some of the meat in human diets (Kupferschmidt, 2015).

The current obesity crisis suggests that food consumption per person is too high in many counties and could be reduced. In recent years the food supply in the US was roughly 4000 calories per capita per day, which is much more than our basic needs, even after taking into account the calories that are wasted (Nestle, 2013, p. xiv). Diets have changed in the past, continue to change in the present, and there is no fundamental reason why they can't change in the future. Combining reductions in consumption with a shift to a more plant-based diet would reduce per-capita food needs and have the added benefit of producing a healthier society.

As much as one-third to one-half of the food produced is not consumed, but is wasted (Foley *et al.*, 2011; Vanham *et al.*, 2015). Much of this waste in developed counties occurs after the food reaches the consumer, while most is lost during harvest, transport and storage in under-developed counties. Regardless of where the loss occurs, reductions in waste are the same as increasing production – more food is available for consumption – and the opportunities to reduce waste are substantial.

A number of unconventional approaches to solving future food problems are being discussed. Vertical farming (growing food in high-rise buildings, abandoned factories, subterranean bomb shelters left from the Second World War) (Despommier, 2009), urban farms that utilize abandoned land (Thomaier *et al.*, 2015), replacement of animal-based foods with similar products synthesized from plant products (e.g. mayonnaise without eggs) (Kowitt, 2014), application of organic farming techniques (Bourne, 2015, pp. 247–271), and growing food close to the point of consumption (locavore movement) (Galzki *et al.*, 2015) are all being touted by enthusiasts as potential solutions to world food problems. Food from these systems is often more expensive and some of them have problems of scale (can producing greens in an abandoned warehouse or in urban gardens in the inner city be scaled up to produce a significant proportion of the food needs of that city?) and they frequently represent replacement of highly efficient, highly productive systems with low-input, low-yield, inefficient systems (Seufert *et al.*, 2012) that usually require larger labour inputs and could require higher energy inputs (vertical farms). The adoption of inefficient production practices is not the best approach to feeding the expanding population, especially when available crop land is limited (Gregory and George, 2011). Buringh and van Heemst (1977, quoted by Evans, 1998, p, 200) estimated that traditional subsistence, labour-oriented agriculture applied to all suitable land area could feed fewer than 4 billion people. The use of systems that produce expensive food in lesser quantities, even if it tastes better and is fresher when it reaches the consumer, does not seem to be the best approach to feeding a world that is getting hungrier. It is unlikely, in my opinion, that these approaches will make a significant contribution to feeding future populations.

The problem of feeding the next 2 billion-plus people added to the world's population seems much more tractable when we realize that there are many options to increase food supplies. One can envision each of these options making a small contribution, with the mix depending on local production systems and the social and economic environment. None of the individual changes would have to be very drastic. Increasing yield will continue to make a contribution, but it won't have to do it all. Cutting back on meat consumption, especially beef, will help, but no one will have to become an absolute vegetarian. Calorie consumption in developed countries could be reduced without leaving us hungry. Waste probably can't be reduced to zero, but it can be reduced, making a significant contribution. Expanding multiple cropping and abandoning biofuel production and other competing land uses would also make a small contribution. When all these relatively small individual contributions are combined, we may find that there is enough food produced in a sustainable manner to feed the extra 2 billion people and the problem is solved.

I am an optimist on the future prospects of feeding the expanding world population, but there are several uncertainties that temper my optimism. These uncertainties will no doubt affect future food production, but the magnitude and the nature of their effect is not yet clear. The first is climate change, driven by the constantly increasing levels of greenhouse gases in the atmosphere. Increasing temperatures will reduce crop yields in some areas, but they may also open up new areas for grain crop production (e.g. the current expansion of maize production into the Prairie Provinces of Canada in North America (Bjerga, 2012)) and increase opportunities for multiple cropping. Rainfall amounts and patterns will probably shift, reducing yields in some areas and increasing them at other locations. It is still difficult to accurately predict changes in precipitation and temperature at the regional level, in spite of extensive climate modelling efforts, so the overall effects on crop production are somewhat uncertain. Evaluations of the effect of the higher temperatures and CO_2 levels expected by 2050 show significant variation among locations, with some crops at some locations showing negligible effects on yield compared with increases or decreases at other crop/location combinations (Hatfield *et al.*, 2011; Fischer *et al.*, 2014, p. 410). Our ability to mitigate the effects of climate change by modifying our crops and cropping systems via plant breeding or management is also an unknown part of the puzzle. The net result of all of these offsetting effects is difficult to determine at present, but many fear that yield growth will ultimately be reduced.

Climate change is also intimately intertwined with the availability of water for irrigation, which is responsible for as much as 34% of the world's agricultural production (Foley *et al.*, 2011). Water supply is a growing concern in many irrigated areas, as ground water is depleted (examples include the Ogallala aquifer in the Midwest of the USA and the Punjab of India) and as reductions in precipitation reduce the availability of surface water (Bourne, 2015, pp. 203–220). These limitations could seriously limit future food supplies and they highlight the tremendous value of areas that have adequate rainfall for high yields without irrigation (examples include central and eastern USA and much of Europe). Some agriculture production may ultimately have to shift from arid to more humid regions.

A second uncertainty is the availability of cheap energy, which is directly related to the issue of greenhouse gas emissions and climate change. Modern agriculture is heavily dependent upon inputs of energy (fuel, fertilizer – especially N, pesticides, machinery), which are mostly supplied by fossil fuels (Gregory and George, 2011). Expensive energy may, at least partially, force the use of lower-input agricultural systems with their attendant reductions in productivity, which could drastically impact our ability to feed 10 billion people. The development of alternative non-CO_2-emitting energy sources may be an important component of the solution to increasing future food supplies.

The potential contribution of molecular biology and genetic engineering to increasing yield represents, in my opinion, another uncertainty. Genetic engineering burst onto the scene in the 1980s with extravagant promises of modified plants that would produce higher yields, be more stress-tolerant and grow with fewer inputs. Thirty-plus years have passed and, to my knowledge, there are no grain crop cultivars engineered specifically to have higher yield in commercial production – they are still in the future. Molecular approaches have created powerful tools to manipulate plants and the fruits of this manipulation, in the form of herbicide and insect resistance, are found in many farmer's fields. It would be foolhardy to suggest that these tools have not and will not contribute to feeding the 10 billion, but, it is not clear what the future contributions will be and when they will be available. Fischer (2011) suggests that contributions from genetic engineering will have to occur quickly, given that the greatest pressure on food supplies will come in the next 20 years, It's possible that 'seeking to explain at the molecular level all trait phenomena may be a costly distraction from seeking to exploit the traits' (Fischer, 2011). Denison (2012, p. 4) expressed concern that excessive spending on biotechnology and consequent underfunding of other areas of research may be counterproductive if biotechnology fails to deliver dramatic yield increases. Whether the approaches of molecular biology will help or hinder efforts to increase yields over the next 35 years remains to be determined.

A final uncertainty is the projected change in population growth rates. World population growth rates are projected to decline by 2050 (United Nations, 2015), reducing the need to increase food production. In Asia and Latin America, growth rates are expected to approach zero or below by 2050, while the population in Europe is projected to decline by 5%; these declines will reduce the pressures on the food supply. Africa represents a huge exception, with the population expected to grow from 1.2 billion today to 3–6 billion by 2100 (Engelman, 2016). Declining growth rates are often associated with improvements in economic conditions, so disruptions of economic growth, that could happen if climate change and changes in energy availability wreak havoc on world economies, could delay the decline in growth rates. Variation in future population growth represents another uncertainty that could influence our efforts to feed the world.

Gloom and doom pronouncements of the future of food sufficiency have been with us at least since the publication of *An Essay on the Principle of Population* by Thomas Robert Malthus in 1798. For example, in 1898, Sir William Crookes predicted mass starvation, as declining levels of soil fertility reduced wheat yields (Crooks, 1898). Jensen (1978) expressed concerns about the continued growth in wheat yields 80 years later. Wennblom (1978) noted in 1978 that crop yields in the Midwestern USA peaked. Paul Ehrlich (1968) and Paddock and Paddock (1967) predicted widespread famine by the 1970s. None of these dire predictions have come to fruition, primarily because technological innovations, unthought of at the time, increased food production and the crisis was averted. In fact, population has more than doubled since 1960 and people are much better fed today than then (Eberstadt, 2006).

The increase in food production since 1798 is truly remarkable. Much of the increase since the middle of the last century was driven by genetic improvements in crop yield and development of crop management practices that helped translate the genetic potential into higher yields in the farmer's fields. Much of the genetic improvement occurred when we had only rudimentary knowledge of the physiology and biochemistry of plant growth and yield production. We now have a much greater understanding of plant growth and much greater capacity to manipulate plants at the genetic level. Whether or not this knowledge will drive yield improvement in the future remains to be determined. Feeding 9.7 billion people by 2050 seems doable when one considers all of the available options to increase the food supply. From this viewpoint, it is hard not to be optimistic about the future. The famous American author William Faulkner put it well when he said: 'I believe that man will not merely endure: he will prevail. He is immortal, not because he alone among creatures has an inexhaustible voice, but because he has a soul, a spirit of compassion and sacrifice and endurance' (Faulkner, 1950).

General Summary

The complexities of the production of food and feed by grain crops can be simplified by grouping the processes involved into two categories – the assimilatory processes occurring primarily in the leaves; and the synthesis processes in the seeds. The assimilatory processes are responsible for the production of the sugars and amino acids that are used by the seeds to produce the carbohydrate, protein and oil that make them useful and valuable. Both groups of processes are integral parts of the production of yield, and one cannot be identified as more fundamental or more important than the other; without either there will be no yield. The focus in this book has been on the latter processes, the accumulation of dry matter by the seed, because it is important and it has not received as much attention in the past as the assimilatory processes.

My objectives in this book, as stated in Chapter 1, were first to develop an understanding of the growth and development of seeds, the processes involved, the regulation of those processes and the effect of plant and environmental factors on them. My second objective was to use this knowledge of seed growth and development to investigate the role of the seed in the yield production process. Accomplishing the first of these objectives produced some relatively simple concepts describing the role of the seed in the determination of yield that greatly expanded our understanding of how yield is produced; an understanding that could not be achieved by considering only the assimilatory parts of the process. What have we learned from these concepts and models describing the role of the individual seed in determining yield?

First, our evaluation identifies canopy photosynthesis during reproductive growth as the primary determinant of yield. Although there have been suggestions that photosynthesis is not related to yield, or that yield is sink-limited, implying that source activity is not important, yield is, in fact, primarily determined by the rate and duration of canopy photosynthesis during reproductive growth. This may seem obvious, but given the complexities of the many plant and environmental factors that affect yield, it's probably worth making the point one more time: photosynthesis and yield are related. High yields require high rates or long durations of canopy photosynthesis, and low yields are usually associated with low rates or short durations of canopy photosynthesis. Conversely, simply increasing the size of the sink will have no effect on yield without an increase in photosynthesis to fill the expanded yield container. I believe that increasing canopy photosynthesis will increase yield. Stresses (water or nutrient stress, disease and insect stress) that commonly reduce yield in the field may have much of their effect by reducing photosynthesis. There are, of course, unique situations where photosynthesis and yield are not related: a sink limitation created by interference of high temperatures with pollination and fertilization is one non-photosynthetic yield-limiting process, and there are others that could be mentioned. We should not, however, extrapolate from these situations, occurring relatively infrequently, to imply that photosynthesis is not important.

Involving the seed in our analysis of yield provides a simple, mechanistic understanding of yield component compensation. We now understand when and why changes in seed size cause compensatory changes in seed number. We can predict, with some confidence, whether or not changes in a yield component will affect yield. This information should remove some of the confusion and mystery from yield component analysis, which should be of some help to crop physiologists because any in-depth consideration of yield must divide it into its components. Regulation of seed growth by the seed provides a framework to analyse potential sink limitations. Quite simply, the sink will be limiting if increases in photosynthesis do not elicit more seeds, or if the individual seed cannot respond to increases in assimilate supply, by growing faster or longer to produce a larger seed.

The analysis of yield production presented in this book doesn't provide all of the answers, and some may argue that more questions are raised than answered. Hopefully, the analysis does provide a useful framework to categorize our knowledge and focus our future research efforts in those areas most important to crop yield and productivity. If this book makes a contribution to this end I will be satisfied.

This entire book has been devoted to gaining a better understanding of the processes involved in the production of yield by grain crops; perhaps it is appropriate to close with some comments about future yields, more specifically, about avenues for future yield improvement. Genetic improvement will probably be the primary vehicle driving future yield gains, especially in highly developed cropping systems. Genetic improvement occurs when defects in the plant are eliminated and/or when the primary productivity of the plant increases (Donald, 1968). The steady improvement of crop plants will make defect elimination increasingly more difficult, so most future yield increases will probably be generated by improvements in primary productivity. It is clear to me that increases in primary productivity will require increases in canopy photosynthesis – either the rate and/or the duration. Canopy photosynthesis produces the bulk of the dry matter accumulated by crops, so the two-to-fivefold increases in yield illustrated in Figs 1.1 and 1.2 must be a result of greater resource capture and an increase in canopy photosynthesis. This increase in canopy photosynthesis occurred by and large without any direct selection for higher photosynthesis, so it was probably a result of many modifications that indirectly affected photosynthesis by, for example, changes in canopy characteristics, reductions in maintenance respiration, increasing stress tolerance or avoidance or insect and disease infestations. Some of these options could be considered defect elimination if one considers susceptibility to stress a defect. Whether more fundamental changes in the photosynthetic apparatus (e.g. modification of Rubisco or inserting the C_4 system into C_3 species) will be involved in the future remains to be determined. Regardless of the cause, higher yields require higher photosynthesis. The increase in photosynthesis will be expressed as more seeds per unit area (most likely) and/or through a longer seed-filling period. Lengthening the duration of photosynthesis and the seed-filling period may be limited by climate and the time available for plant growth, but this restriction can be accommodated by shortening the vegetative growth period to allow a longer seed-filling period without lengthening the total duration.

The goal is clear, but the best approach to achieve the goal is still being debated. Currently, much effort seems to be focused on the utilization of specific yield-related traits (Reynolds *et al.*, 2009; Marte *et al.*, 2015), implying that some version of the ideotype approach popularized by C.M. Donald (Donald, 1968) is superior to simply selecting for yield. As discussed earlier in this chapter, and given the need for fairly immediate improvements in yield to meet food demands in the next 20 to 30 years, I believe that maintaining strong traditional breeding efforts, where selection is focused on yield, is the most prudent approach. Advanced technologies developed by molecular biologists could supplement the traditional approach to increase its efficiency. The traditional breeding approach has been very successful, driving yields of all grain crops steadily upward since the middle of the 20th century. The ideotype approach does not have such a track record to recommend it. Molecular biology has produced a quantum increase in our ability to manipulate plant processes since the publication of the first edition of this book in 1998, but, in spite of this great potential, this approach to date has made only limited contributions to increasing the yield of grain crops. Given the complexities of the yield production process and the lack of proven success of trait-based, ideotype and molecular approaches, direct selection for yield is still the best approach in the short run; putting resources into other approaches may be too much of a gamble. In a perfect world with unlimited funding, both approaches could proceed simultaneously, but that perfect world does not exist, so I prefer taking the conservative approach that has successfully fed the world for the past 75 years or so.

References

Abernathy, R.H., Palmer, R.G., Shibles, R. and Anderson, J.C. (1977). Histological observations on abscising and retained soybean flowers. *Canadian Journal of Plant Science* 57, 713–716.

Ackerson, R.C., Havelka, V.D. and Boyle, M.G. (1984). CO_2 enrichment effects on soybean physiology. II. Effects of stage-specific CO_2 exposure. *Crop Science* 24, 1150–1154.

Acreche, M.M. and Slafer, G.A. (2006). Grain weight response to increases in number of grains of wheat in a Mediterranean area. *Field Crops Research* 98, 52–59.

Adams, C.A. and Rinne, R.W. (1980). Moisture content as a controlling factor in seed development and germination. *International Review of Cytology* 68, 1–8.

Adams, M.W. (1967). Basis of yield component compensation in crop plants with special reference to the field bean, *Phaseolus vulgaris*. *Crop Science* 7, 505–510.

Afuakwa, J.J. and Crookston, R.K. (1984). Using the kernel milk line to visually monitor grain maturity in maize. *Crop Science* 24, 687–691.

Alberta Agriculture. (2015). Triticale grain for feed – nutritional information. Available at: http://www1.agric.gov.ab.ca/$department/deptdocs.nsf/all/fcd10575 (accessed 17 June 2016).

Ahmadi, A. and Baker, D.N. (2001). The effect of water stress on grain filling processes in winter wheat. *Journal of Agriculture Science* 136, 257–269.

Aldrich, S.R. (1943). Maturity measurements in corn and an indication that grain development continues after premature cutting. *Journal of the American Society of Agronomy* 35, 667–688.

Allen, D.K. and Young, J.D. (2013). Carbon and nitrogen provisions alter the metabolic flux in developing soybean embryos. *Plant Physiology* 161, 1458–1475.

Alves, A.A.C. (2002). Cassava: botany and physiology. In: Hillocks, R.J., Thresh, J.M. and Bellotti, A.C. (eds) *Cassava: Biology, Production and Utilization*. CAB International, Wallingford, Oxfordshire, UK, pp. 67–89.

Anderson, S.R. (1955). Development of pods and seeds of birdsfoot trefoil, *Lotus corniculatus* L., as related to maturity and to seed yields. *Agronomy Journal* 47, 483–487.

Andrade, F.H. and Ferreiro, M.A. (1996). Reproductive growth of maize, sunflower and soybean at different source levels during grain filling. *Field Crops Research* 48, 155–165.

Andrade, F.H., Uhart, S.A. and Frugone, M.I. (1993). Intercepted radiation at flowering and kernel number in maize: Shade versus plant density effects. *Crop Science* 33, 482–485.

Andrade, F.H., Cirilo, A.G. and Echarte, L. (2000). Factors affecting kernel number in maize. In: Otegui, M.E. and Slafer, G.A. (eds) *Physiological Bases for Maize Improvement*. Food Products Press, New York, USA, pp. 59–74.

Andrade, F.H., Calvino, P., Cirilo A. and Barbieri, P. (2002). Yield response to narrow rows depend on increased radiation interception. *Agronomy Journal* 94, 975–980.

Andrade, F.H., Sadras, V.O., Vega, C.R.C. and Echarte, L. (2005). Physiological determinants of crop growth and yield in maize, sunflower, and soybean. Their application to crop management, modeling and breeding. *Journal of Crop Improvement* 14, 51–101.

Andrade, F.H., Poggio, S.L., Ermacora, M. and Satorre, E.H. (2015). Productivity and resource use in intensified cropping systems in the Rolling Pampa, Argentina. *European Journal of Agronomy* 67, 37–51.

Andrade-Sanchez, P., Gore, M.A., Heun, J.T., Thorp, K.R., Cormo, A.E., French, A.N., Salvucci, M.E. and White, J.W. (2014). Development and evaluation of a field-based high-throughput phenotyping platform. *Functional Plant Biology* 41, 68–79.

Aparicio-Tejo, P.M. and Boyer, J.S. (1983). Significance of accelerated leaf senescence at low water potentials for water loss and grain yield in maize. *Crop Science* 23, 1198–1202.

Araki, H., Hamada, A., Hossain, A. and Takahashi, T. (2012). Waterlogging at jointing and/or anthesis induces early leaf senescence and impairs grain filling. *Field Crops Research* 137.

Araus, J.L., Slafer, G.A., Reynolds, M.P. and Royo, C. (2002). Plant breeding and drought in C$_3$ cereals: What should we breed for? *Annuals of Botany* 89, 925–940.

Artlip, T.S., Madison, J.T. and Setter, T.L. (1995). Water deficit in developing endosperm of maize: Cell division and nuclear DNA endoreduplication. *Plant, Cell and Environment* 18, 1034–1040.

Ashley, D.A. and Counce, P.A. (1993). Weight loss during maturation in cereal grains. *Agronomy Abstracts* 149.

ASPCA (2016). Shelter intake and surrender. American Society for the Prevention of Cruelty to Animals. Available at: http://www.aspca.org/animal-homelessness/shelter-intake-and-surrender/pet-statistics (accessed 27 August 2016).

Aspinall, D. (1965). The effects of soil moisture stress on the growth of barley. II. Grain growth. *Australian Journal of Agriculture Research* 16, 265–275.

Asseng, S., Cao, W., Zhang, W. and Ludwig, F. (2009). Crop physiology modeling and climate change: Impact and adaptation strategies. In: Sadras V.O. and Calderini D.E. (eds) *Crop Physiology. Applications for Genetic Improvement and Agronomy.* Academic Press, London, UK, pp. 511–543.

Austin, R.B. (1994). Plant breeding opportunities. In: Boote, K.J., Bennett, J.M., Sinclair, T.R. and Paulsen, G.M. (eds) *Physiology and Determination of Crop Yield* ASA-CSSA-SSSA, Madison, Wisconsin, USA, pp. 567–586.

Austin, R.B., Ford, M.A., Morgan, C.L. and Yeoman, D. (1993). Old and modern wheat cultivars compared on the Broadbalk wheat experiment. *European Journal of Agronomy* 141–147.

Azam-Ali, S.N. and Squire, G.R. (2002). *Principles of Tropical Agronomy*, CAB International, Wallingford, Oxfordshire, UK.

Ball, R.A., McNew, R.W., Vories, E.D., Keisling, T.C. and Purcell, L.C. (2001). Path analysis of population density effects on short–season soybean yield. *Agronomy Journal* 93, 187–195.

Banziger, M., Feil, B. and Stamp, P. (1994). Competition between nitrogen accumulation and grain growth for carbohydrates during grain filling of wheat. *Crop Science* 34, 440–446.

Barlow, E.W.R., Donovan, G.R. and Lee, J.W. (1983). Water relations and composition of wheat ears grown in liquid culture: Effect of carbon and nitrogen. *Australian Journal of Plant Physiology* 10, 99–108.

Bassetti, P. and Westgate, M.E. (1993a). Emergence, elongation and senescence of maize silks. *Crop Science* 33, 271–275.

Bassetti, P. and Westgate, M.E. (1993b). Water deficit affects receptivity of maize silks. *Crop Science* 33, 279–282.

Bassi, F.M., Bently, A.R., Charmet, G., Ortiz, R. and Crossa, J. (2016). Breeding schemes for the implementation of genomic selection in wheat (*Triticum spp.*). *Plant Science* 242, 23–36.

Bastidas, A.M., Setiyono, T.D., Dobermann, A., Cassman, K.G., Elmore, R.W., Graef, G.L. and Specht, J.E. (2008). Soybean sowing date: The vegetative, reproductive and agronomic impacts. *Crop Science* 48, 727–740.

Beddow, J.M. and Pardley, P.G. (2015). Moving matters: The effect of location on crop production. *Journal of Economic History* 75, 219–249.

Belamkar, V., Wenger, A., Kalberer, S.R., Bhattacharya, V.G., Blackmon, W.J. and Cannon, S.B. (2015). Evaluation of phenotypic variation in a collection of *Apios Americana*: An edible tuberous legume. *Crop Science* 55, 712–726.

Below, F.E., Christensen, A.J., Reed, A.J. and Hageman, R H. (1981). Availability of reduced N and carbohydrates for ear development of maize. *Plant Physiology* 68: 1186–1190.

Bewley, J.D. and Black, M. (1994). *Seeds: Physiology of Development and Germination*, 2nd edn. Plenum Press, New York, USA.

Bewley, J.D., Bradford, K.J., Hilhorst, H.W.M. and Nonogak,i H. (2013). *Seeds. Physiology of Development, Germination and Dormancy*, 3rd edn. Springer, New York, USA.

Bhullar, S.S. and Jenner, C.F. (1986). Effects of temperature on the conversion of sucrose to starch in the developing wheat endosperm. *Australian Journal of Plant Physiology* 13, 605–615.

Bieler, P., Fussell, L.K. and Bidinger, F.R. (1993). Grain growth of *Pennisetum glaucum* (L.). *Field Crops Research* 31, 41–45.

Biello, D. (2011). The false promise of biofuels. *Scientific American* 305, 59–65.

Bils, R.F. and Howell, R.W. (1963). Biochemical and cytological changes in developing soybean cotyledons. *Crop Science* 3, 304–308.

Bishnoi, U.R. (1974). Physiological maturity of seeds in *Triticale hexaploid* L. *Crop Science* 14, 819–821.

Bittman, M. (2014). Don't ask how to feed the 9 billion. *New York Times*, 11 November, A27. Available at: http://www.nytimes.com/2014/11/12/opinion/dont-ask-how-to-feed-the-9-billion.html (accessed 31 August 2016).

Bjerga, A. (2012). Canada's new corn belt attracts hot money to bargain farmland. Bloomberg News. Available at: http://www.farmlandgrab.org/post/view/21281 (accessed 31 August 2016).

Black, M., Bewley, J.D. and Halmer, P. (2006). *The Encyclopedia of Seeds: Science, Technology and Uses*. CAB International, Wallingford, Oxfordshire, UK.

Blacklow, W.M. and Incoll, L.D. (1981). Nitrogen stress of winter wheat changed the determinants of yield and the distribution of nitrogen and total dry matter during grain filling. *Australian Journal of Plant Physiology* 8, 191–200.

Blackman, V.H. (1919). The compound interest law and plant growth. *Annals of Botany* 33, 353–360.

Blum, A. (2009). Effective use of water (EUW) and not water use efficiency (WUE) is the target of crop yield improvement under drought stress. *Field Crops Research* 112, 119–123.

Blum, A. and Arki, G.F. (1984). Sorghum root growth and water–use as affected by water supply and growth duration. *Field Crops Research* 9, 131–142.

Board, J.E. and Tan, Q. (1995). Assimilatory capacity effects on soybean yield components and pod number. *Crop Science* 35, 846–851.

Boerma, H.R. and Ashley, D.A. (1988). Canopy photosynthesis and seed fill duration in recently developed soybean cultivars and selected plant introductions. *Crop Science* 28, 137–140.

Bolanos, J. (1995). Physiological basis for yield differences in selected maize cultivars from Central America. *Field Crops Research* 42, 69–80.

Bolanos, J. and Edmeades, G.O. (1996). The importance of the anthesis–silking interval in breeding for drought resistance in tropical maize. *Field Crops Research* 48, 65–80.

Boon-Long, P., Egli, D.B. and Leggett, J.E. (1983). Leaf N and photosynthesis during re-productive growth in soybeans. *Crop Science* 23, 617–620.

Boote, K.J. and Sinclair, T.R. (2006). Crop physiology: Significant discoveries and our changing perspective on research. *Crop Science* 46, 2270–2277.

Boote, K.J., Jones, J.W. and Pickering, N. (1996). Potential uses and limitations of crop models. *Agronomy Journal* 88, 704–716.

Boote, K.J., Jones, J.W., Hoogenboom, G. and Pickering, N. (1998). The CROPGRO model for grain legumes. In: Tsuji G.Y., Hoogenboom G. and Thorton P.K. (eds) *Understanding Options for Agricultural Production*. Kluwar Academic Dordrecht. The Netherlands.

Borras, L. and Gambin, B.L. (2010). Trait dissection of maize kernel weight: Towards in-tegrating hierarchical scales using a plant growth approach. *Field Crops Research* 148, 1–12.

Borras, L. and Westgate, M.E. (2006). Predicting maize kernel sink capacity early in devel-opment. *Field Crops Research* 95, 223–233.

Borras, L., Westgate, M.E. and Otegui M.E. (2003). Control of kernel weight and kernel water relations by post-flowering source–sink ratio in maize. *Annals of Botany* 91, 857–867.

Borras, L., Slafer, G.A. and Otegui, M.E. (2004). Seed dry weight response to source–sink manipulations in wheat, maize and soybean: A quantitative reappraisal. *Field Crops Research* 86, 131–146.

Borrell, A.K. and Hammer, G.L. (2000). Nitrogen dynamics and physiological basis for stay-green in sorghum. *Crop Science* 40, 1295–1307.

Bort, J., Brown, R.H. and Araus, J.L. (1996). Refixation of respiratory CO_2 in the ears of C3 cereals. *Journal of Experimental Botany* 47, 1567–1575.

Boshankian, S. (1918). The mechanical factors determining the shape of the wheat kernel. *Journal of the American Society of Agronomy* 10, 205–209.

Bourne, J.K.J. (2015). *The End of Plenty. The Race to Feed a Crowded World.* W.W. Norton and Co., New York, USA.

Bowers, G.R. (1995). An early soybean production system for drought avoidance. *Journal of Production Agriculture* 8, 112–119.

Boyer, J.S. and Mclaughlin, J.E. (2007). Functional reversion to identify controlling genes in multigenic responses: Analysis of floral abortion. *Journal of Experimental Botany* 58, 267–277.

Bradford, K.J. (1994). Water stress and the water relations of seed development: A critical review. *Crop Science* 34, 1–10.

Bremner, P.M. (1969). Effects of time and rate of nitrogen application on tillering, 'sharp eyespot' (*Rhizoctonia solani*) and yield in winter wheat. *Journal of Agricultural Science* 72, 273–280.

Brevedan, R.E. and Egli, D.B. (2003). Short periods of water stress during seed filling, leaf senescence and yield of soybean. *Crop Science* 43, 2083–2088.

Brevedan, R.E., Egli, D.B. and Leggett, J.E. (1978). Influence of N nutrition on flower and pod abortion and yield of soybeans. *Agronomy Journal* 70, 81–84.

Briaty, L.G., Coult, D.A. and Boulter, D. (1969). Protein bodies of developing seeds of *Vicia faba*. *Journal of Experimental Botany* 20, 358–372.

Briggs, G.E., Kidd, F. and West, C. (1920). A quantitative analysis of plant growth: Part II. *Annals of Applied Biology* 7, 202–223.

Brim, C.A. and Burton, J.W. (1979). Recurrent selection in soybean. II. Selection for increased percent protein in seeds. *Crop Science* 19, 494–498.

Brisson, N., Gate, P., Gouache, D., Charmet, G., Oury, F.–X. and Huard, F. (2010). Why are wheat yields stagnating in Europe? A comprehensive data analysis for France. *Field Crops Research* 119, 201–212.

Brocklehurst, P.A. (1977). Factors controlling grain weight in wheat. *Nature* 266, 348–349.

Brooking, I.R. (1990). Maize ear moisture during grain-filling and its relation to physiological maturity and grain-drying. *Field Crops Research* 23, 55–68.

Brooks, A., Jenner, C.F. and Aspinall, D. (1982). Effects of water deficit on endosperm starch granules and on grain physiology of wheat and barley. *Australian Journal of Plant Physiology* 9, 423–436.

Brown, R.J., Cruse, R.M. and Colvin, T.S. (1989). Tillage system effects on crop growth and production costs for a corn–soybean rotation. *Journal of Production Agriculture* 2, 273–279.

Browne, C.L. (1978). Identification of physiological maturity in sunflowers (*Helianthus annus*). *Australian Journal of Experimental Agriculture and Animal Husbandry* 18, 282–286.

Bruening, W.P. and Egli, D.B. (1999). Relationship between photosynthesis and seed number at phloem isolated nodes in soybean. *Crop Science*. 39, 1769–1775.

Bruening, W.P. and Egli, D.B. (2003). Leaf starch accumulation and seed set at phloem-isolated nodes in soybean. *Field Crops Research* 68, 113–120.

Bunting A.H. (1971). Productivity and profit or Is your vegetative phase really necessary? *Annals of Applied Biology* 67, 265–285.

Bunting, A.H. (1975). Time, phenology and the yields of crops. *Weather* 30, 312–325.

Buringh, P. and van Heemst, H.D.J. (1977). *An Estimation of World Food Production Based on Labour-oriented Agriculture*. Centre for World Food Market Research, Amsterdam, The Netherlands.

Calderini, D.F. and Reynolds, M.P. (2000). Changes in grain weight as a consequence of de-graining treatments at pre- and post-anthesis in a synthetic hexaploid wheat (*Triticum durum* x *T, taushchi*). *Australian Journal of Plant Physiology* 27, 183–191.

Calderini, D.F., Abeledo, L.G., Savin, R. and Slafer, G.A. (1999). Effect of temperature and carpel size during pre-anthesis on potential grain weight in wheat. *Journal of Agricultural Science* 132, 453–460.

Calderini, D.E., Abeledo, L.G. and Slafer, G.A. (2000). Physiological maturity in wheat based on kernel water and dry matter. *Agronomy Journal* 92, 895–901.

Caley, C.Y., Duffus, C.M. and Jeffcoat, B. (1990). Photosynthesis in the pericarp of developing wheat grains. *Journal of Experimental Botany* 41, 303–308.

Campbell, J.E., Lobell, D.B., Genova, R.C. and Field C.B. (2008). The global potential of bioenergy on abandoned agriculture lands. *Environmental Science and Technology* 42, 5791–5794.

Cantagallo, J.E., Chimente, C.A. and Hall, A.J. (1997). Number of seed per unit area in sunflower correlates well with a photothermal quotient. *Crop Science* 37, 1780–1786.

Capitanio, R., Gentinetta, E. and Motto M. (1983). Grain weight and its components in maize inbred lines. *Maydica* 28, 365–379.

Carceller, J.L. and Aussenac, T. (1999). Accumulation and changes in molecular size distribution of polymeric proteins in developing grains of hexaploid wheats: Role of the desiccation phase. *Australian Journal of Plant Physiology* 26, 301–310.

Carcova, J.M., Uribelarrea, L., Borras, L., Otegui, M.E. and Westgate, M.E. (2000). Synchronous pollination within and between ears improves kernel set in maize. *Crop Science* 40, 1056–1061.

Carlson, J.B. and Lersten, N.R. (1987). Reproductive morphology. In: Wilcox, J.R. (ed.) *Soybeans: Improvement, Production and Uses*, 2nd edn. Agronomy Monograph 16, Crop Science Society of America, Madison, Wisconsin, USA, pp. 95–134.

Carr, D.J. and Skene, K.G.M. (1961). Diauxic growth curves of seeds, with special reference to French beans (*Phaseolus vulgaris* L.). *Australian Journal of Biological Science* 14, 1–12.

Carter, M.W. and Poneleit, C.G. (1973). Black layer maturity and filling period variation among inbred lines of corn (*Zea mays* L.). *Crop Science* 13, 436–439.

Cassman, K.G. (1999). Ecological intensification of cereal productions systems: Yield potential, soil quality and precision agriculture *Proceedings of the National Academy of Sciences* 96, 5952–5959.

Cassman, K.G., Grassini, P. and van Wart, J. (2010). Crop yield potential, yield trends, and global food security in a changing climate. In: Rosenzweig, C. and Hillel, D. (eds) *Handbook of Climate Change and Agroecosystems*. Imperial College Press, London, UK, pp. 37–51.

Cavalieri, A.J. and Smith, O.S. (1985). Grain filling and field drying of a set of maize hybrids released from 1930 to 1982. *Crop Science* 25, 856–861.

Caviglia, O.P., Sadras, V.O. and Andrade, F.H. (2004). Intensification of agriculture in the south-eastern Pampas. I. Capture and efficiency in the use of water and radiation in double-cropped wheat–soybean. *Field Crops Research* 87, 117–129.

Chandler, R.F.J. (1969). Improving the rice plant and its culture. *Nature* 221, 1007–1010.

Chapman, S.C. and Edmeades, G.O. (1999). Selection improves drought tolerance in tropical maize populations. II. Direct and correlated responses among secondary traits. *Crop Science* 39, 1315–1324.

Charles-Edwards, D.A. (1982). *Physiological Determinants of Crop Growth*. Academic Press Australia, North Ryde, New South Wales, Australia.

Charles-Edwards, D.A. (1984a). On the ordered development of plants. 1. An hypothesis. *Annals of Botany* 53, 699–707.

Charles-Edwards, D.A. (1984b). On the ordered development of plants. 2. Self-thinning in plant communities. *Annals of Botany* 53, 709–714.

Charles-Edwards, D.A. and Beech, D.F. (1984). On the ordered development of plants. 3. Branching by the grain legume *Cyamopsis tetragonoloba* (Guar). *Annals of Botany* 54, 673–679.

Charles-Edwards, D.A., Doley, D. and Rimmington, G.M. (1986). *Modelling Plant Growth and Development*. Academic Press Australia, Sydney, Australia

Chen, C., Qian, C., Peng, A. and Zhang, W. (2012). Progressive and active adaption of cropping systems to climate change in Northeast China. *European Journal of Agronomy* 38, 94–103.

Chimenti, C.A., Hall, A.J. and Sol Lopez, M. (2001). Embryo-growth rate and duration in sunflower as affected by temperature. *Field Crops Research* 69, 81–88.

Chojecki, A.J.S., Bayliss, M.W. and Gale, M.D. (1986). Cell production and DNA accumulation in the wheat endosperm and their association with grain weight. *Annals of Botany* 58, 809–817.

Chowdhury, S.I. and Wardlaw, I.F. (1978). The effect of temperature on kernel development in cereals. *Australian Journal of Agricultural Research* 29, 205–223.

Christy, A.L. and Porter, C.A. (1982). Canopy photosynthesis and yield in soybean. In: Govindjee (ed.) *Photosynthesis, Development, Carbon Metabolism and Plant Productivity*. Academic Press, New York, USA, pp. 499–511.

Cirilo, A.G. and Andrade, F.H. (1994). Sowing date and maize productivity: II. Kernel number determination. *Crop Science* 34, 1044–1046.

Clarke, J.M. (1983). Time of physiological maturity and post-physiological maturity drying rates in wheat. *Crop Science* 23, 1203–1205.

Classen, M.M. and Shaw, R.H. (1970). Water deficit effects on corn. II. Grain components. *Agronomy Journal* 62, 652–655.

Cobb, G., Hole, B.D.J., Smith, J.D. and Kent, M.W. (1988). The effects of modifying sucrose concentration on the development of maize kernels grown *in vitro*. *Annals of Botany* 62, 265–270.

Cochrane, M.P. and Duffus, C.M. (1983). Endosperm cell number in cultivars of barley differing in grain weight. *Annals of Applied Biology* 102, 177–181.

Cock, J.H. and Yoshida, S. (1973). Changing sink and source relations in rice (*Oryza sativa* L.) using carbon dioxide enrichment in the field. *Soil Science and Plant Nutrition* 19, 229–234.

Cohen, J.E. (1995). *How Many People Can the Earth Support?* W.W. Norton, New York, USA.

Cooper, P.J.M. (1979). The association between altitude, environmental variables, maize growth and yields in Kenya. *Journal of Agriculture Science* 93, 635–649.

Copeland, L.O. and McDonald, M.B. (2001). *Principles of Seed Science and Technology*. Klumar Academic Publishers, Boston, Massachusetts, USA.

Copeland, P.J. and Crookston, R.K. (1985). Visual indicators of physiological maturity. *Crop Science* 25, 843–847.

Corner, E.J.H. (1951). The leguminous seed. *Phytomorphology* 1, 117–150.

Coste, F., Ney, B. and Crozat, Y. (2001). Seed development and seed physiological quality of field grown beans (*Phaseolus vulgaris* L.). *Seed Science and Technology* 29, 121–136.

CRA (2016). NASA Prediction of Worldwide Energy Resource (POWER). Climatology Resource for Agroclimatology. Available at: http://power.larc.nasa.gov/cgi-bin/cgiwrap/solar/agro.cgi (accessed 27 August 2016).

Crafts-Brandner, S.J. and Egli, D.B. (1987). Sink removal and leaf senescence in soybean. *Plant Physiology* 85, 662–666.

Crafts-Brandner, S.J. and Poneleit, C.G. (1992). Selection for seed growth characteristics: Effect on leaf senescence in maize. *Crop Science* 32, 127–131.

Crain, J.L., Wei, Y., Barker, J., Thompson, S.M., Alderman, P.D., Reynolds, M., Zhang, N. and Poland, J. (2016). Development and deployment of a portable field phenotyping platform. *Crop Science* 56, 965–975.

Crooks, W. (1898). Address of the president before the British Association for the Advancement of Science, 1898. *Science* 8, 561–575.

Crookston, R.K. and Hill, D.S. (1978). A visual indicator of the physiological maturity of soybean seed. *Crop Science* 18, 867–870.

Crosbie, T.M. and Mock, J.J. (1981). Changes in physiological traits associated with grain yield improvement in three maize breeding programs. *Crop Science* 21, 255–259.

Cross, H.Z. (1975). Diallel analysis of duration and rate of grain filling of seven inbred lines of corn. *Crop Science* 15, 532–535.

Cruz-Aguado, J.A., Reyes, F. and Dorado, M. (1999). Effect of source-to-sink ratios on partitioning of dry matter and 14C–photoassimilates in wheat during grain filling. *Annals of Botany* 83, 655–665.

Cully, D.E., Gengenbach, B.G., Smith, J.A., Rubenstein, I., Connelly, J.A. and Park, W.O. (1984). Endosperm protein synthesis and L–(^{35}S) methionine incorporation in maize kernels cultured *in vitro*. *Plant Physiology* 74, 389–384.

Curtis, D.L. (1968). The relation between the date of heading of Nigerian sorghums and the duration of the growing season. *Journal of Applied Ecology* 5, 215–226.

Dapaah, H.K., McKenzie, B.A. and Hill, G.D. (2000). Influence of sowing date and irrigation on growth and yield of Pinto beans (*Phaseolus vulgarius*) in a sub-humid temperate environment. *Journal of Agricultural Science* 134, 33–43.

Dardanelli, J.L., Bachneir, O.A., Sereno, R. and Gil, R. (1997). Rooting depth and soil water extraction patterns of different crops in a silty loam Haplustoll. *Field Crops Research* 54, 29–38.

Darrock, B.A. and Baker, R.J. (1995). Two measures of grain filling in spring wheat. *Crop Science* 35, 164–168.

Davies, D.R. (1975). Studies of seed development in *Pisum sativum* 1. Seed size in reciprocal crosses. *Planta* 124, 297–302.

Davies, D.R. (1977). DNA contents and cell number in relation to seed size in the genus *Vica*. *Heredity* 39, 153–164.

Davies, S., Turner, N.C., Siddique, K.H.M., Plummer, J. and Leport, L. (1999). Seed growth of desi and kabuli chickpea (*Cicer arietinum* L.) in a short season Mediterranean-type environment. *Australian Journal of Experimental Agriculture* 39, 181–188.

Daynard, T.B. (1972). Relationships among black layer formation, grain moisture percentage and heat unit accumulation in corn. *Agronomy Journal* 64, 716–719.

Daynard, T.B. and Duncan, W.G. (1969). The black layer and grain maturity in corn. *Crop Science* 9, 473–476.

Daynard, T.B. and Kannenberg, L.W. (1976). Relationships between length of the actual and effective grain filling periods and the grain yield of corn. *Canadian Journal of Plant Science* 56, 237–242.

Daynard, T.B., Tanner, J.W. and Duncan, W.G. (1971). Duration of the grain filling period and its relationship to grain yield in corn, *Zea mays* L. *Crop Science* 11, 45–48.

de Souza, P.I., Egli, D.B. and Bruening, W.P. (1997). Water stress during seed filling and leaf senescence in soybean. *Agronomy Journal* 89, 807–812.

de Wit, C.T. (1965). *Photosynthesis of leaf canopies*. Institute for Biological and Chemical Research on Field Crops and Herbs, Wageningen, The Netherlands.

de Wit, C.T. (1967). Photosynthesis: its relationship to overpopulation. In: SanPierto, A., Green, F.A. and Army, T.J. (eds) *Harvesting the Sun*. Academic Press, New York, USA, pp. 315–320.

DeKhuijzen, H.M. and Verkerke, D.R. (1986). The effect of temperature on development and dry-matter accumulation of *Vicia faba* seeds. *Annals of Botany* 58, 869–885.

Denison, R.F. (2012). *Darwinian Agriculture. How Understanding Evolution Can Improve Agriculture*. Princeton University Press, Princeton, New Jersey, USA.

Despommier, D. (2009). The rise of vertical farms. *Scientific American* 301, 80–87.

Dimmock, J.P.R.E. and Gooding, M.J. (2002). The effects of fungicides on rate and duration of grain filling in winter wheat in relation to maintenance of flag leaf green area. *Journal of Agricultural Science* 138, 1–16.

Djarot, J.N. and Peterson, D.M. (1991). Seed development in a shrunken endosperm barley mutant. *Annals of Botany* 68, 495–499.

Doehlert, D.C., Jannik, J.–L. and McMullen, M.S. (2008). Size distribution of different orders of kernels within the oat spikelet. *Crop Science* 48, 298–304.

Dofing, S.M. (1997). Ontogentic evaluation of grain yield and time to maturity in barley. *Agronomy Journal* 89, 685–690.

Dominguez, C. and Hume, D.J. (1978). Flowering, abortion, and yield of early-maturing soybeans at three densities. *Agronomy Journal* 70, 801–805.

Donald, C.M. (1962). In search of yield. *Journal of Australian Institute of Agricultural Science* 28, 171–178.

Donald, C.M. (1968). The breeding of crop ideotypes. *Euphytica* 17, 385–403.

Donald, C.M. and Hamblin, J. (1976). The biological yield and harvest index of cereals as agronomic and plant breeding criteria. *Advances in Agronomy* 28, 361–405.

Donovan, G.R., Lee, J.W., Longhurst, T.J. and Martin, P. (1983). Effect of temperature on grain growth and protein accumulation in cultured wheat ears. *Australian Journal of Plant Physiology* 10, 445–450.

Duncan, W.G. (1967). Model building in photosynthesis. In: Pietro, A.S. and Greer, F.A. (eds) *Harvesting the Sun – Photosynthesis in Plant Life*, Academic Press, New York, USA, pp. 309–313.

Duncan, W.G. (1969). Cultural manipulation for higher yields. In: Eastin, J.D., Haskins, F.A., Sullivan, C.Y. and van Bavel, C.H.M. (eds) *Physiological Aspects of Crop Yield*. ASA-CSSA-SSSA, Madison, Wisconsin, USA, pp. 327–339.

Duncan, W.G. (1971). Leaf angles, leaf area, and canopy photosynthesis. *Crop Science* 11, 482–485.

Duncan W.G. (1975). Maize. In: Evans, L.T. (ed.) *Crop Physiology*. Cambridge University Press, London, UK, pp. 23–50.

Duncan, W.G., Loomis, R.S., Williams, W.A. and Hanau, R. (1967). A model simulating photosynthesis in plant communities. *Hilgardia* 38, 181–205.

Duncan, W.G., Shaver, D.L. and Williams, W.A. (1973). Insolation and temperature effects on maize growth and yield. *Crop Science* 13, 187–189.

Duncan, W.G., McCloud, D.E., McGraw, R.L. and Boote, K.J. (1978). Physiological basis for yield improvement in peanut. *Crop Science* 18, 1015–1020.

Dunphy, E.J., Hanway, J.J. and Green, D.E. (1979). Soybean yields in relation to days between specific developmental stages. *Agronomy Journal* 71, 917—920.

Dunwell, J.M. (2010). Crop biotechnology: Prospects and opportunities. *Journal of Agriculture Science* 149, 17–27.

Duthion, C. and Pigeaire, A. (1991). Seed lengths corresponding to the final stage in seed abortion in three grain legumes. *Crop Science* 31, 1597–1583.

Duvick, D.N. (1984). Genetic contributions to yield gains of U.S. hybrid maize, 1930 to 1980. In: Fehr, W.R. (ed.) *Genetic Contributions to Yield Gains of Five Major Crop Plants*. Spec. Pub. No. 7, Crop Science Society of America, Madison, Wisconsin, USA, pp. 15–45.

Duvick, D.N. (1992). Genetic contributions to advances in yield of U.S. maize. *Maydica* 37, 69–79.

Duvick, D.N. (2005). The contribution of breeding to yield advances in maize (*Zea mays*). *Advances in Agronomy* 86, 84–145.

Duvick, D.N. and Cassman, K.G. (1999). Post-green revolution trends in yield potential of temperate maize in the North–central United States. *Crop Science* 39, 1622–1630.

Dwyer, L.M., Ma, B.L., Everson, L. and Hamilton, R.I. (1994). Maize physiological traits related to grain yield and harvest moisture. *Crop Science* 34, 985–992.

Easterling, D.R. (2002). Recent changes in frost days and the frost-free season in the United States. *Bulletin of the American Meteorological Society* 83, 1327–1332.

Eastin, J.D., Hultquist, J.H. and Sullivan, C.Y. (1973). Physiologic maturity in grain sorghum. *Crop Science* 13, 175–17

Eastmond, P., Kolacna, L. and Rawsthorne, S. (1996). Photosynthesis by developing embryos of oilseed rape (*Brassica napus* L.). *Journal of Experimental Botany* 47, 1763–1769.

Eberstadt, N. (2006). Doom and demography. *The Wilson Quarterly* 30, 27–31.

Echarte, L., Andrade, F.H., Vega, C.R.C. and Tollenaar ,M. (2000). Kernel determination in Argentinean hybrids released between 1965 and 1993. *Crop Science* 44, 1654–1661.

Echarte, M.M., Alberdi, I. and Aguirrezabal, L.A.N. (2012). Post-flowering assimilate availability regulates oil fatty acid composition in sunflower grains. *Crop Science* 52, 818–829.

Edwards, J.T. and Purcell, L.C. (2005a). Soybean yield and biomass responses to increasing plant population among diverse maturity groups. I. Agronomic characteristics. *Crop Science* 45, 1770–1777.

Edwards, J.T. and Purcell, L.C. (2005b). Light interception and yield response of ultra-short-season soybean to diphenylether herbicides in the Midsouthern USA. *Weed Technology*. 19, 168–175.

Edwards, J.T., Purcell, L.C., Vories, E.D., Shannon, J.G. and Ashlock, L.O. (2003). Short-season soybean cultivars have similar yields with less irrigation than longer season cultivars. *Crop Management* 2(1). DOI: 10.1094/cm–2003–0922–01–RS.

Edwards, J.T., Purcell, L.C. and Vories, E.D. (2005). Light interception and yield potential of short-season maize (*Zea maize* L.) hybrids in the midsouth. *Agronomy Journal* 97, 225–234.

Egli, D.B. (1981). Species differences in seed growth characteristics. *Field Crops Research* 4, 1–12.

Egli, D.B. (1990). Seed water relations and the regulation of the duration of seed growth in soybean. *Journal of Experimental Botany* 41, 243–248.

Egli, D.B. (1993). Cultivar maturity and potential yield of soybean. *Field Crops Research* 32, 147–158.

Egli, D.B. (1994a). Cultivar maturity and reproductive growth duration in soybean. *Journal of Agronomy and Crop Science* 173, 249–254.

Egli, D.B. (1994b). Mechanisms responsible for soybean yield response to equidistant planting patterns. *Agronomy Journal* 86, 1046–1049.

Egli, D.B. (1997). Cultivar maturity and response of soybean to shade stress during seed filling. *Field Crops Research* 52, 1–8.

Egli, D.B. (1998). *Seed Biology and the Yield of Grain Crops*. CAB International, Wallingford, Oxfordshire, UK.

Egli, D.B. (1999). Variation in leaf starch and sink limitations during seed filling in soybean. *Crop Science* 39, 1361–1368.

Egli, D.B. (2004). Seed-fill duration and yield of grain crops. *Advances in Agronomy* 83, 243–279.

Egli, D.B. (2008). Comparison of corn and soybean yield trends in the United States: Historical trends and future prospects. *Agronomy Journal* 100, S79–S88.

Egli, D.B. (2010). Soybean reproductive sink size and short-term reductions in photosynthesis during flowering and pod set. *Crop Science* 50, 1971–1977.

Egli, D.B. (2011). Time and the productivity of agronomic crops and cropping systems *Agronomy Journal* 103, 743–750.

Egli, D.B. (2012). Timing of fruit initiation and seed size in soybean. *Journal of Crop Improvement* 26, 751–766.

Egli, D.B. (2013). The relationship between the number of nodes and pods in soybean communities. *Crop Science* 53, 1668–1676.

Egli, D.B. (2015a). Pod set in soybean: Investigations with SOYPODP, a whole plant model. *Agronomy Journal* 107, 349–360.

Egli, D.B. (2015b). Is there a role for sink size in understanding maize population–yield relationships? *Crop Science* 55, 1–10.

Egli, D.B. and Bruening, W.P. (2000). Potential of early-maturing soybean cultivars in late plantings. *Agronomy Journal* 92, 532–537.

Egli, D.B. and Bruening, W.P. (2001). Source–sink relationships, seed sucrose levels and seed growth rates in soybean. *Annals of Botany* 88, 235–242.

Egli, D.B. and Bruening, W.P. (2002). Synchronous flowering and fruit set at phloem-isolated nodes in soybean. *Crop Science* 42, 1535–1540.

Egli, D.B. and Bruening, W.P. (2003). Increasing sink size does not increase photosynthesis during seed filling in soybean. *European Journal of Agronomy* 19, 289–298.

Egli, D.B. and Bruening, W.P. (2004). Water stress, photosynthesis, seed sucrose levels and seed growth in soybean. *Journal of Agricultural Science* 142, 1–8.

Egli, D.B. and Bruening, W.P. (2005). Shade and temporal distribution of pod production and pod set in soybean. *Crop Science* 45, 1764–1769.

Egli, D.B. and Bruening, W.P. (2006a). Temporal patterns of pod production and pod set in soybean. *European Journal of Agronomy* 24, 11–18.

Egli, D.B. and Bruening, W.P. (2006b). Fruit development and reproductive survival: Position and age effects. *Field Crops Research* 98, 195–202.

Egli, D.B. and Cornelius, P.L. (2009). A regional analysis of the response of soybean yield to planting date. *Agronomy Journal* 101, 330–335.

Egli, D.B. and Hatfield, J.L. (2014a). Yield gaps and yield relationships in central U.S. soybean production systems. *Agronomy Journal* 106, 560–566.

Egli, D.B. and Hatfield, J.L. (2014b). Yield and yield gaps in central U.S. corn production systems. *Agronomy Journal* 106, 2248–2254.

Egli, D.B. and Leggett, J.E. (1973). Dry matter accumulation patterns in determinate and indeterminate soybeans. *Crop Science* 13, 220–222.

Egli, D.B. and Leggett, J.E. (1976). Rate of dry matter accumulation in soybean seeds with varying source–sink ratios. *Agronomy Journal* 68, 371–374.

Egli, D.B. and TeKrony, D.M. (1997). Species differences in seed water status during seed maturation and germination. *Seed Science Research* 7, 3–11.

Egli, D.B. and Wardlaw, I.F. (1980). Temperature response of seed growth characteristics of soybean. *Agronomy Journal* 72, 560–564.

Egli, D.B. and Zhen-wen, Y. (1991). Crop growth rate and seed number per unit area in soybean. *Crop Science* 31, 439–442.

Egli, D.B., Leggett, J.E. and Wood, J.M. (1978a). Influence of soybean seed size and position on the rate and duration of filling. *Agronomy Journal* 70, 127–130.

Egli, D.B., Leggett, J.E. and Duncan, W.G. (1978b). Influence of N stress on leaf senescence and N redistribution in soybean. *Agronomy Journal* 70, 43–47.

Egli, D.B., Leggett, J.E. and Cheniae, A. (1980). Carbohydrate levels in soybean leaves during reproductive growth. *Crop Science* 20, 468–473.

Egli, D.B., Fraser, J., Leggett, J.E. and Poneleit, C.G. (1981). Control of seed growth in soya beans (*Glycine max* (L.) Merrill). *Annals of Botany* 48, 171–176.

Egli, D.B., Guffy, R.D., Meckel, L.W. and Leggett, J.E. (1985a). The effect of source–sink alterations on soybean seed growth. *Annals of Botany* 55, 395–402.

Egli, D.B., Guffy, R.D. and Leggett, J.E. (1985b). Partitioning of assimilate between vegetative and reproductive growth in soybean. *Agronomy Journal* 77, 917–922.

Egli, D.B., Duncan, W.G. and Crafts-Brandner, S.J. (1987a). Effect of physical restraint on seed growth in soybean. *Crop Science* 27, 289–294.

Egli, D.B., Guffy, R.D. and Heitholt, J.J. (1987b). Factors associated with reduced yields of delayed plantings of soybean. *Journal of Agronomy and Crop Science* 159, 176–185.

Egli, D.B., Wiralaga, R.A. and Ramseur, E.L. (1987c). Variation in seed size in soybean. *Agronomy Journal* 79, 463–467.

Egli, D.B., Wiralaga, R.A., Bustamam, T., Yu, Z.-W. and TeKrony, D.M. (1987d). Time of flower opening and seed mass in soybean. *Agronomy Journal* 79, 697–700.

Egli, D.B., Ramseur, E.L., Zhen-wen, Y. and Sullivan, C.H. (1989). Source–sink alterations affect the number of cells in soybean cotyledons. *Crop Science* 29, 732–735.

Egli, D.B., TeKrony, D.M. and Wiralaga, R.A. (1990). Effect of soybean seed vigor and size on seedling growth. *Journal of Seed Technology* 14, 1–12.

Ehlers, J.D. and Hall, A.E. (1997). Cowpea (*Vigna unguiculata* L. Walp). *Field Crops Research* 53, 187–204.

Ehrlich, P.R. (1968). *The Population Bomb*. Ballantine Books, New York, USA.

Ellis, R.H. and Pieta-Filho, C. (1992). The development of seed quality in spring and winter cultivars of barley and wheat. *Seed Science Research* 2, 9–15.

Engelman, R. (2016). Six billion in Africa. *Scientific American* 39, 56–63.

Engledow, F.L. and Wadham, S.M. (1923). Investigations of yield in cereals I. *Journal of Agriculture Science* 13, 390–439.

Evans, L.T. (1975). Crops and the world food supply, crop evolution and the origins of crop physiology. In: Evans, L.T. (ed.) *Crop Physiology: Some Case Histories*. Cambridge University Press, Cambridge, UK.

Evans, L.T. (1993). *Crop Evolution, Adaptation and Yield*. Cambridge University Press, Cambridge, UK.

Evans, L.T. (1998). *Feeding the Ten Billion. Plants and Population Growth*. Cambridge University Press, Cambridge, UK.

Evans, L.T., Bingham, J. and Roskams, M.A. (1972). The pattern of grain set within ears of wheat. *Australian Journal of Biology* 25, 1–8.

Evans, L.T. and Dunstone, R.L. (1970). Some physiological aspects of evolution in wheat. *Australian Journal of Biological Science* 23, 725–741.

Evans, L.T., Wardlaw, I.F. and Fischer R.A. (1975). Wheat. In: Evans, L.T. (ed.) *Crop Physiology: Some Case Histories*. Cambridge University Press, Cambridge, UK, pp. 101–149.

Fageria, N.K. (2007). Yield physiology of rice. *Journal of Plant Nutrition* 30, 843–879.

Fageria, N.K., Baligar, V.C. and Clark, R.B. (2006). *Physiology of Crop Production*. Food Products Press, Binghamton, New York, USA.

Fakorede, M.A.B. and Mock, J.J. (1978). Changes in morphological and physiological traits associated with recurrent selection for grain yield in maize. *Euphytica* 27, 397–409.

FAOSTAT (2016). Crop production statistics. Available at: http://faostat.fao.org/site/567/DesktopDefault.aspx (accessed 1 September 2016).

Faulkner, W. (1950) William Faulkner – banquet speech, 10 December. Available at: http://www.nobelprize.org/nobel_prizes/literature/laureates/1949/faulkner-speech.html (accessed 31 August 2016).

Fehr, W.R. and Caviness, C.E. (1977). *Stages of Soybean Development*. Special Report 80. Iowa State University, Ames, Iowa, USA.

Fischer, K.S. and Palmer, A.F.E. (1984). Tropical maize. In: Goldsworthy, P.R. and Fisher, N.M. (eds) *The Physiology of Tropical Field Crops*. John Wiley and Sons, New York, USA, pp. 213–248.

Fischer, R.A. (1975). Yield potential in a dwarf spring wheat and the effect of shading. *Crop Science* 15, 607–613.

Fischer, R.A. (1979). Growth and water limitation to dryland wheat yield in Australia: A physiological framework. *Journal of Australian Institute of Agricultural Science* 45, 83–94.

Fischer, R.A. (1985). Number of kernels in wheat crops and the influence of solar radiation and temperature. *Journal of Agricultural Science, Cambridge* 105, 447–461.

Fischer, R.A. (2011). Wheat physiology: A review of recent developments. *Crop and Pasture Science* 62, 95–114.

Fischer, R.A. and Aguilar, I.M. (1976). Yield potential in a dwarf spring wheat and the effect of carbon dioxide fertilization. *Agronomy Journal* 68, 749–752.

Fischer, R.A. and Kohn, G.D. (1966). The relationship of grain yield to vegetative growth and post-flowering leaf area in the wheat crop under conditions of limited soil moisture. *Australian Journal of Agricultural Research* 17, 281–295.

Fischer, R.A. and Laing, D.R. (1976). Yield potential in a dwarf spring wheat and response to crop thinning. *Journal of Agricultural Science, Cambridge* 87, 113–122.

Fischer, R.A. and Stockman, J.M. (1980). Kernel number per spike in wheat: Responses to pre-anthesis shading. *Australian Journal of Plant Physiology* 7, 169–180.

Fischer, R.A., Aguilar, I. and Laing, D.R. (1977). Post-anthesis sink size in a high-yielding dwarf wheat: yield response to grain numbers. *Australian Journal of Agricultural Research* 28, 165–175.

Fischer, T., Byerlee, D. and Edmeades, G. (2014). *Crop Yields and Global Food Security.* A CIAR monograph, No. 158. Australian Centre for International Agricultural Research, Canberra, Australia.

Fisher, M. (2014). B.B. Singh and his quest to make cowpea the food legume of the 21st century. *CSA News*, October, 4–10. Available at: https://dl.sciencesocieties.org/publications/csa/pdfs/59/10/4 (accessed 31 August 2016).

Fisher, D.B. and Gifford, R.M. (1986). Accumulation and conversion of sugars by developing wheat grains. VI. Gradients along the transport pathway from the peduncle to the endosperm cavity during grain filling. *Plant Physiology* 82, 1024–1030.

Foley, J.A., Ramankutty, N., Brauman, K.A., Cassidy, E.S., Gerber, J.S., Johnson, M., Mueller, N.D., O'Connell, C., Deepak, K.R., West, P.C. *et al.* (2011). Solutions for a cultivated planet. *Nature* 478, 337–342.

Ford, D.M., Shibles, R.M. and Green, D.E. (1983). Growth and yield of soybean lines selected for divergent leaf photosynthesis ability. *Crop Science* 23, 517–520.

Fortescue, J.A. and Turner, D.W. (2007). Changes in seed size and oil accumulation in *Brassica napus* L. by manipulating the source–sink ratio and excluding light from developing siliques. *Australian Journal of Agricultural Research* 58, 413–424.

Foulkes, M.J., Reynolds, M.J. and Sylvester-Bradley, R. (2009). Genetic improvement of grain crops yield potential. In: Sadras, V.O. and Calderine, D.F. (eds) *Crop Physiology. Applications for Genetic Improvement and Agronomy.* Academic Press, San Diego, California, USA.

Frank, S.J. and Fehr, W.R. (1981). Associations among pod dimensions and seed weight in soybeans. *Crop Science* 21, 547–550.

Fraser, J., Egli, D.B. and Leggett, J.E. (1982a). Pod and seed development in soybean cultivars with differences is seed size. *Agronomy Journal* 74, 81–85.

Fraser, J., Dougherty, C.T. and Langer, R.H.M. (1982b). Dynamics of tiller populations of standard height and semi-dwarf wheats. *New Zealand Journal of Agriculture Research* 25, 321–328.

Frederick, J.R. and Camberato, J.J. (1995). Water and nitrogen effects on winter wheat in the southeastern coastal plain I. Grain yield and kernel traits. *Agronomy Journal* 87, 521–526.

Frederick, J.R., Hesketh, J.D., Peters, D.B. and Below, F.E. (1989). Yield and reproductive trait response of maize hybrids to drought stress. *Maydica* 34, 118–122.

Freier, G., Vilella, F. and Hall, A.J. (1984). Within-ear pollination synchrony and kernel set in maize. *Maydica* 29, 317–324.

Frey, N.M. (1981). Dry matter accumulation in kernels of maize. *Crop Science* 21, 118–122.

Frey, K.J., Ruan, E. and Wiggans, S.C. (1958). Dry weights and germination of developing oat seeds. *Agronomy Journal* 50, 248–250.

Fujita, K., Coronel, V.P. and Yoshida, S. (1984). Grain-filling characteristics of rice varieties (*Oryza sativa* L.) differing in grain size under controlled environmental conditions. *Soil Science and Plant Nutrition* 30, 445–454.

Fukumorita, T. and Chino, M. (1982). Sugar, amino acids and inorganic contents of rice phloem. *Plant Cell Physiology* 23, 273–283.

Furbank, R.T., White, R., Palta, J.A. and Turner, N.C. (2004). Internal recycling of respiratory CO_2 in pods of chickpea (*Cicer arietinum* L.): the role of pod wall, seed coat and embryo. *Journal of Experimental Botany* 55, 1687–1696.

Fussell, L.K. and Dwarte, D.M. (1980). Structural changes of the grain associated with black region formation in *Pennisetum americanum*. *Journal of Experimental Botany* 31, 645–654.

Fussell, L.K. and Pearson, C.J. (1978). Course of grain development and its relationship to black region appearance in *Pennisetum americanum*. *Field Crops Research* 1, 21–31.

Gaju, O., Reynolds, M.P., Sparkes, D.L. and Foulkes, M.J. (2009). Relationship between large-spike phenotype, grain number and yield potential in spring wheat. *Crop Science* 49, 961–973.

Galzki, J.C., Mula, D.J. and Peters, C.J. (2015). Mapping the potential of local food capacity in Southwestern Minnesota. *Renewable Agriculture and Food Systems* 30, 364–372.

Gambin, B.L. and Borras, L. (2005). Sorghum kernel weight: Growth patterns from different positions within the panicle. *Crop Science* 45, 553–561.

Gambin, B.L. and Borras, L. (2007). Plasticity of sorghum kernel weight to increase assimilate availability. *Field Crops Research* 100, 272–284.

Gambin, B.L., Borras, L. and Otegui, M.E. (2007). Kernel water relations and duration of grain filling in maize temperate hybrids. *Field Crops Research* 101, 1–9.

Gambin, B.L., Borras, L. and Otegui, M.E. (2008). Kernel weight dependence upon plant growth at different grain-filling stages in maize and sorghum. *Australian Journal of Agricultural Research* 59, 280–290.

Gao, X., Francis, D., Ormrod, J.C. and Bennett, M.D. (1993). Changes in cell number and cell division activity during endosperm development in allohexaploid wheat, *Triticum aestivum*. *Journal of Experimental Botany* 43, 1063–1610.

Garcia del Moral, L.F., Ramos, J.M., Garcia del Moral, M.B. and Jumenez-Tejada, M.P. (1991). Ontogenic approach to grain production in spring barley based on path-coefficient analysis. *Crop Science* 31, 1179–1185.

Gardner, F.P., Valle, R. and McCloud, D.E. (1990). Yield characteristics of ancient races of maize compared to modern hybrids. *Agronomy Journal* 82, 864–868.

Gay, S., Egli, D.B. and Reicosky, D.A. (1980). Physiological aspects of yield improvement in soybeans. *Agronomy Journal* 72, 387–391.

Gbikpi, P.J. and Crookston, R.K. (1981). Effect of flowering date on accumulation of dry matter and protein in soybean seeds. *Crop Science* 21, 652–655.

Gebeyehou, G., Knott, D.R. and Baker, R.J. (1982). Relationships among durations of vegetative and grain filling phases, yield components and grain yield in durum wheat cultivars. *Crop Science* 22, 287–290.

Gelinas, B. and Seguin, P. (2008). Evaluation of management practices for grain Amaranth production in Eastern Canada. *Agronomy Journal* 100, 344–351.

Gent, M.P.N. (1995). Canopy light interception, gas exchange, and biomass in reduced height isolines of winter wheat. *Crop Science* 35, 1636–1642.

Gerik, T.J., Rosenthal, W.D., Vanderlip, R.L. and Wade, L.J. (2004). Simulating seed number in grain sorghum from increases in plant dry weight. *Agronomy Journal* 96, 1222–1230.

Gesch, R.W. and Johnson, B.L. (2012). Seed moisture at physiological maturity in oil seed and confectionary sunflower hybrids in the northern U.S. *Field Crops Research* 133, 1–9.

Gibson, L.R. and Mullins, R.E. (1996). Influence of day and night temperatures on soybean seed yield. *Crop Science* 36, 98–104.

Gifford, R.M. and Thorne, J.H. (1985). Sucrose concentration at the apoplastic interface between seed coat and cotyledons of developing soybean seeds. *Plant Physiology* 77, 863–868.

Gimenez, C. and Fereres, E. (1986). Genetic variability in sunflower cultivars under drought. II. Growth and water relations. *Australian Journal Agriculture Research* 37, 583–597.

Glover, J.D., Reganold, J.P., Bell, L.W., Boreuitz, J., Brummer, E.C., Buckler, E.S., Cox, C.M., Cox, J.S., Crews, T.E., Culman, S.W. *et al.* (2010). Increased food and ecosystem security via perennial grains. *Science* 328, 1638–1639.

Goldsworthy, P.R. (1984). Crop growth and development: The reproductive phase. In: Goldsworthy, P.R. and Fisher, N.M. (eds) *The Physiology of Tropical Field Crops.* John Wiley & Sons Ltd, New York, USA, pp. 163–212.

Goldsworthy, P.R. and Colegrove, M. (1974). Growth and yield of highland maize in Mexico. *Journal of Agricultural Science Cambridge* 83, 213–221.

Grabau, L.J., Van Sandford, D.A. and Meng, Q.W. (1990). Reproductive characteristics of winter wheat subjected to postanthesis shading. *Crop Science* 30, 771–774.

Grabe, D.F. (1989). Measurement of seed moisture. In: Standwood, P.C. and McDonald, M.B. (eds) *Seed Moisture.* CSSA Special Publication 14. Crop Science Society of America, Madison, Wisconsin, USA, pp. 69–92.

Grafius, J.E. (1978). Multiple characters and correlated response. *Crop Science* 18, 931–934.

Graham, P.H. and Ranalli, P. (1997). Common bean (*Phaseolus vulgaris* L.). *Field Crops Research* 53, 131–146.

Grassini, P., Eskridge, K.M. and Cassman, K.G. (2013). Distinguishing between yield advances and yield plateaus in historical crop production trends. *Nature Communications* 4. Article no. 2918. DOI: 10.1038/ncomms3918.

Gregory, P.J. and George, T.S. (2011). Feeding the nine billion: the challenge to sustainable crop production. *Journal of Experimental Botany* 62, 5233–5239.

Grist, D.H. (1986). *Rice.* Longman Group Limited, New York, USA.

Guldan, S.J. and Brun, W.A. (1985). Relationship of cotyledon cell number and seed respiration to soybean seed growth. *Crop Science* 25, 815–819.

Hall, A.E. (1999). Cowpea. In: Smith, D.L. and Hamel, C. (eds) *Crop Yield: Physiology and Processes.* Springer-Verlag, Berlin, Germany, pp. 355–373.

Hall, C.A. and Kitgaard, K.A. (2012). *Energy and the Wealth of Nations. Understanding the Biophysical Economy.* Springer, New York, USA.

Hall, A.J. and Richards, R.A. (2013). Prognosis for genetic improvement of yield potential and water-limited yield of major grain crops *Field Crops Research* 143, 18–33.

Hall, A.J., Lemcoff, J.H. and Trapani, N. (1981). Water stress before and during flowering in maize and its effects on yield, its components and their determinants. *Maydica* 26, 19–38.

Hall, A.J., Conner, D.J. and Whitfield, D.M. (1989). Contribution of pre-anthesis assimilate to grain filling in irrigated and water stressed sunflower. I. Estimates using labeled carbon. *Field Crops Research* 20, 95–112.

Hall, A.J., Whitfield, D., M. and Conner, D.J. (1990). Contribution of pre-anthesis assimilates to grain–filling in irrigated and water-stressed sunflower crops. II. Estimates from a carbon budget. *Field Crops Research* 24, 273–294.

Hallauer, A.R. and Russell, W.R. (1962). Estimates of maturity and its inheritance in maize. *Crop Science* 2, 289–294.

Hamid, Z.A. and Grafius, J.E. (1978). Developmental allometry and its implications for grain yield in barley. *Crop Science* 18, 83–86.

Han, T., Wu, C., Tong, Z., Mentreddy, R.S., Tan, K. and Gai, J. (2006). Post flowering photoperiod regulates vegetative growth and reproductive development of soybean. *Environmental and Experimental Botany* 55, 120–129.

Hanft, J.M. and Wych, R.D. (1982). Visual indicators of physiological maturity of hard red spring wheat. *Crop Science* 22, 584–588.

Hanft, J.M., Jones, R.J. and Stumme, A.B. (1986). Dry matter accumulation and carbohydrate concentration patterns of field grown and in vitro cultured maize kernels from tip and middle ear positions. *Crop Science* 26, 568–572.

Hanway, J.J. and Weber, C.R. (1971). Dry matter accumulation in eight soybean (*Glycine max* L. Merrill) varieties. *Agronomy Journal* 63, 227–230.

Hardman, L.L. and Brun, W.A. (1971). Effects of atmospheric carbon dioxide enrichment at different development stages on growth and yield components of soybeans. *Crop Science* 11, 886–888.

Harlan, H.V. (1920). Daily development of kernels of Haunchen barley from flowering to maturity at Aberdeen, Idaho. *Journal of Agricultural Research* 19, 393–430.

Harlan, J.R. (1992). *Crops and Man*, 2nd edn. Crop Science Society of America, Madison, Wisconsin, USA.

Harlan, H.V. and Pope, M.N. (1923). Water content of barley kernels during growth and maturation. *Journal of Agriculture Research* 23, 333–360.

Harrington, J.F. (1972). Seed storage and longevity. In: Kozlowski, T.T. (ed.) *Seed Biology*, Vol III. Academic Press. New York, USA, pp. 145–245.

Harris, R.E., Moll, R.H. and Stuber, C.W. (1976). Control and inheritance of prolificacy in maize. *Crop Science* 16, 843–850.

Hartung, R.C., Poneleit, C.G. and Cornelius, P.L. (1989). Direct and correlated responses to selection for rate and duration of grain fill in maize. *Crop Science* 29, 740–745.

Hartwig, E.E. and Edwards, C.J. (1970). Effects of morphological characteristics upon seed yield in soybeans. *Agronomy Journal* 62, 64–65.

Hashemi-Dezfouli, A. and Herbert, S.J. (1992). Intensifying plant density response of corn with artificial shade. *Agronomy Journal* 84, 547–551.

Hatfield, A.L. and Ragland, J.L. (1967). New concepts in corn growth. *Plant Food Review* 12, 1–3.

Hatfield, J.L. and Walthall, C.L. (2015). Meeting global food needs: Realizing the potential via genetics x environment x management interactions. *Agronomy Journal* 107, 1215–1226.

Hatfield, J.L., Boote, K.J., Kimball, B.A., Ziska, L.H., Izaurralde, R.C., Ort, D., Thomson, A.M. and Wolfe, D. (2011). Climate impacts on agriculture: Implications for crop production. *Agronomy Journal* 103, 351–370.

Hawkins, R.C. and Cooper, P.J.M. (1981). Growth, development and grain yield of maize. *Experimental Agriculture* 17, 203–207.

Hay, R.K.M. (1995). Harvest index: A review of its use in plant breeding and crop physiology. *Annals of Applied Biology* 126, 197–216.

Hay, R.K.M. and Kirby, E.J.M. (1991). Convergence and synchrony – a review of the coordination of development in wheat. *Australian Journal of Agricultural Research* 42, 661–700.

Hay, R.K.M. and Porter, J.R. (2006). *The Physiology of Crop Yield*, 2nd edn. Blackwell Publishing Ltd, Oxford, UK.

Hayashi, H. and Chino, M. (1986). Collection of pure phloem sap from wheat and its chemical composition. *Plant Cell Physiology* 27, 1387–1393.

Hayati, R., Egli, D.B. and Crafts-Brandner, S.J. (1995). Carbon and nitrogen supply during seed filling and leaf senescence in soybean. *Crop Science* 35, 1063–1069.

Hayati, R., Egli, D.B. and Crafts-Brandner, S.J. (1996). Independence of nitrogen supply and seed growth in soybean: studies using an *in vitro* culture system. *Journal of Experimental Botany* 47, 33–40.

Heatherly, L.G. (1999). Early soybean production system (ESPS). In: Heatherly, L.G. and Hodges, H.F. (eds) *Soybean Production in the Midsouth*. CRC Press, Boca Raton, Florida, USA, pp. 103–118.

Heatherly, L.G. and Elmore, R.W. (2004). Managing inputs for peak production. In: Boerma, H.R. and Specht, J.E. (eds) *Soybeans: Improvement, Production, and Uses*, 3rd edn. ASA-CSSA-SSSA, Madison, Wisconsin, USA, pp. 451–536.

Heinrich, G.M., Francis, C.A., Eastin, J.D. and Saeed, M. (1985). Mechanisms of yield stability in sorghum. *Crop Science* 25, 1109–1113.

Heiser, C.B.J. (1973). *Seed to Civilization – The Story of Man's Food*. W.H. Freeman & Co., San Francisco, California, USA.

Heitholt, J.J., Egli, D.B. and Leggett, J.E. (1986). Characteristics of reproductive abortion in soybean. *Crop Science* 26, 589–595.

Heitholt, J.J., Croy, L.I., Maness, N.O. and Nguyen, H.T. (1990). Nitrogen partitioning in genotypes of winter wheat differing in kernel N concentration. *Field Crops Research* 23, 133–144.

Hellewell, K.B., Stuthman, D.D., Markhart, A.H. and Erwin, J.E. (1996). Day and night temperature effects during grain-filling in oat. *Crop Science* 36, 624–628.

Helsel, D.B. and Frey, K.J. (1978). Grain yield variation in oats associated with differences in leaf area duration among oat lines. *Crop Science* 18, 765–769.

Herbert, S.J. and Litchfield, G.U. (1984). Growth response of short season soybean to variations in row spacing and density. *Field Crops Research* 9, 163–171.

Herrero, M.P. and Johnson, R.R. (1980). High temperature stress and pollen viability of maize. *Crop Science* 20, 796–800.

Hesketh, J.D. and Moss, D.N. (1963). Variation in the response of photosynthesis to light. *Crop Science* 3, 107–110.

Hesketh, J.D., Myhre, D.L. and Willey C.R. (1973). Temperature control of time intervals between vegetative and reproductive events in soybeans. *Crop Science* 13, 250–254.

Housley, T.L., Kirleis, A.W., Ohm, H.W. and Patterson, F.L. (1982). Dry matter accumulation in soft red winter wheat seeds. *Crop Science* 22, 290–294.

Howell, R.W., Collins, F.I. and Sedgewick, V.E. (1959). Respiration of soybean seed as related to weathering losses during ripening. *Agronomy Journal* 51, 677–679.

Hsu, F.C., Bennett, A.B. and Spanswick, R.M. (1984). Concentrations of sucrose and nitrogenous compounds in the apoplast of developing soybean seed coats and embryos. *Plant Physiology* 75, 181–186.

Huber, A.G., and Grabe, D.F. (1987). Endosperm morphogenesis in wheat: Termination of nuclear division. *Crop Science* 27, 1252–1256.

Huff, A. and Dybing, C.D. (1980). Factors affecting shedding of flowers in soybean *Glycine max* (L.) Merrill. *Journal of Experimental Botany* 31, 751–762.

Hulse J.H., Laing E.M. and Pearson O.E. (1980). *Sorghum and the Millets: Their Composition and Nutritive Value*. Academic Press, London, UK.

Hunt, R. (1978). *Plant Growth Analysis*. The Institute of Biology's Studies in Biology, No. 96. Edward Arnold, London, UK.

Hunt, L.A., van der Poorten, G. and Pararajiasingham, S. (1991). Postanthesis temperature effects on duration and rate of grain filling in some winter and spring wheats. *Canadian Journal of Plant Science* 71, 609–617.

Hunter, J.L., TeKrony, D.M., Miles, D.F. and Egli, D.B. (1991). Corn seed maturity indicators and their relationship to uptake of carbon-14 assimilate. *Crop Science* 31, 1309–1313.

Husain, M.M., Hill, G.D. and Gallagher, J.N. (1988). The response of field bean (*Vicia faba* L.) to irrigation and sowing time I. Yield and yield components. *Journal of Agricultural Science* 111, 221–232.

Ibrahim, A.E., TeKrony, D.M., Egli, D.B. and van Sanford D.A. (1992). Water content and germination of immature wheat kernels. *Seed Science and Technology* 20, 39–46.

Inoue, H. and Tanaka, A. (1978). Comparison of source and sink potentials between wild and cultivated potatoes. *Journal of Science of Soil and Manure Japan* 49, 321–328.

Ishibashi, T., Sneller, S.C. and Shannon, J.G. (2003). Soybean yield potential and phenology in the ultra-short season production system. *Agronomy Journal* 95, 1082–1087.

Islam, M.S. and Morison, J.I.L. (1992). Influence of solar radiation and temperature on irrigated rice grain yield in Bangladesh. *Field Crops Research* 30, 13–28.

Jacobs, B.C. and Pearson, C.J. (1991). Potential yield of maize determined by rates of growth and development of ears. *Field Crops Research* 27, 281–289.

Jamboonsri, W., Phillips, T., Geneve, R., Cahill, J. and Hildebrand, D. (2012). Extending the range of an ancient crop, *Salvia hispanica* – a new ω3 source. *Genetic Resources and Crop Evolution* 59, 171–178.

Jenner, C.F. (1980). Effects of shading or removing spikelets in wheat: testing assumptions. *Australian Journal of Plant Physiology* 7, 113–121.

Jenner, C.F. and Rathjen, A.J. (1977). Supply of sucrose and its metabolism in developing grains of wheat. *Australian Journal of Plant Physiology* 4, 691–701.

Jenner, C.F. and Rathjen, A.J. (1978). Physiological basis of genetic differences in growth of grains of six wheat varieties. *Australian Journal of Plant Physiology* 5, 249–262.

Jenner, C.F., Ugalde, T.D. and Aspinall, D. (1991). The physiology of starch and protein deposition in the endosperm of wheat. *Australian Journal of Plant Physiology* 18, 211–226.

Jensen, N.F. (1978). Limits to growth in world food production. *Science* 201, 317–320.

Jeuffroy, M.-H. and Bouchard, C. (1999). Intensity and duration of nitrogen deficiency on wheat grain number. *Crop Science* 39, 1385–1353.

Jeuffroy, M.-H. and Ney, B. (1997). Crop physiology and productivity. *Field Crops Research* 53, 3–16.

Jiang, H. (1993). The relationship between photosynthesis during flowering and pod set and seeds per unit area in soybean. MS thesis. University of Kentucky, Lexington, Kentucky, USA.

Jiang, H. and Egli, D.B. (1993). Shade induced changes in flower and pod number and fruit abscission in soybean. *Agronomy Journal* 85, 221–225.

Jiang, H. and Egli, D.B. (1995). Soybean seed number and crop growth rate during flowering. *Agronomy Journal* 87, 264–267.

Johann, H. (1935). The histology of the caryopsis of yellow dent corn with reference to resistance and susceptibility to kernel rots. *Journal of Agriculture Research* 51, 855–883.

Johnson, D.R. and Tanner, J.W. (1972). Comparisons of corn (*Zea mays* L.) inbreds and hybrids grown at equal leaf area index, light penetration, and population. *Crop Science* 12, 482–485.

Jones R.J., and Setter, T.L. (2000). Hormonal regulation of early kernel development. In: Westgate, M.E. and Boote, K.J. (eds) *Physiology and Modeling of Kernel Set in Maize.* Crop Science Society of America. Madison, Wisconsin USA, pp. 25–42.

Jones, R.J. and Simmons, S.R. (1983). Effect of altered source–sink ratio on growth of maize kernels. *Crop Science* 23, 129–135.

Jones, D.B., Peterson, M.L. and Geng, S. (1979). Association between grain filling and yield components in rice. *Crop Science* 19, 641–643.

Jones, R.J., Gengenbach, B.G. and Cardwell, V.B. (1981). Temperature effect on in–vitro kernel development in maize. *Crop Science* 21, 761–766.

Jones, R.J., Quattar, S. and Crookston, R.K. (1984). Thermal environment during endosperm cell division and grain filling in maize: Effects on kernel growth and development *in vitro*. *Crop Science* 24, 133–137.

Jones, R.J., Roessler, J. and Quattar, S. (1985). Thermal environment during endosperm cell division in maize: Effects on the number of endosperm cells and starch granules. *Crop Science* 25, 830–834.

Jones, R.J., Schreiber, B.M.N. and Rossler, J.A. (1996). Kernel sink capacity in maize: genotypic and maternal regulation. *Crop Science* 36, 301–306.

Jones, J.W., Hoogenboom, G., Porter, C.H., Boote, K.J., Batchelor, D., Hunt, L.A., Wilkens, P.W., Singh, U., Gijsman, A.J. and Ritchie, J.T. (2003). The DSSAT cropping system model. *European Journal of Agronomy* 18, 235–265.

Jongkaeuwattana, S., Geng, S., Hill, J.E. and Miller, B.C. (1993). Within-panicle variability of grain filling in rice cultivars with different maturities. *Journal of Agronomy and Crop Science* 171, 236–242.

Jurgens, S.K., Johnston, R.R. and Boyer, J.S. (1978). Dry matter production and translocation in maize subjected to drought during grain filling. *Agronomy Journal* 70, 678–682.

Kamoshita, A., Fukai, S., Muchow, R.C. and Cooper, M. (1998). Sorghum hybrid differences in grain yield and nitrogen concentration under low soil nitrogen availability. I. Hybrids with similar phenology. *Australian Journal of Agricultural Research* 49, 1267–1276.

Kantolic, A.G. and Slafer, G.A. (2001). Photoperiod sensitivity after flowering and seed number determination in indeterminate soybean cultivars. *Field Crops Research* 72, 109–118.

Karlen, D.L., Duffy, M.D. and Colvin, T.S. (1995). Nutrients, labor, energy and economic evaluation of two farming systems in Iowa. *Journal of Production Agriculture* 8, 540–546.

Kato, T. (1986). Effects of shading and rachis–branch clipping on the grain–filling process of rice cultivars differing in the grain size. *Japanese Journal of Crop Science* 55, 252–260.

Kato T. (1999). Genetic and environmental variations and associations of the characters related to the grain–filling process in rice. *Plant Production Science* 2, 32–36.

Kiesselbach, T.A. (1949). *The Structure and Reproduction of Corn.* Research Bulletin 161. University of Nebraska, Lincoln, Nebraska USA.

Kiesselbach, T.A. and Walker, E.R. (1952). Structure of certain specialized tissues in the kernel of corn. *American Journal of Botany* 39, 561–569.

Kiniry, J.R. (1988). Kernel weight increase in response to decreased kernel number in sorghum. *Agronomy Journal* 80, 221–222.

Kiniry, J.R. and Knievel, D.P. (1995). Response of maize and seed number to solar radiation intercepted soon after anthesis *Agronomy Journal* 87, 228–234.

Kiniry, J.R., Spanel, D.A. and Beckholt, A.J. (1990). Seed weight response to seed number in maize. *Agronomy Journal* 82, 98–102.

Klinkhamer, P.G.L., Meeks, E., de Jong, T.J. and Weiner, J. (1992). On the analysis of size-dependent reproductive output in plants. *Functional Ecology* 6, 308–316.

Kobata, T., Yoshida, H., Masiko, U. and Honda, T. (2013). Spikelet sterility is associated with a lack of assimilate in high-spikelet-number rice. *Agronomy Journal* 105, 1821–1831.

Kole, S. and Gupta, K. (1982). The timing of physiological maturity of seeds in sunflower: evaluation through multiple tests. *Seed Science and Technology* 10, 457–467.

Koller, H.R., Nyquist, W.E. and Chorush, I.S. (1970). Growth analysis of the soybean community. *Crop Science* 10, 407–412.

Kolloffel, C. and Matthews, S. (1983). Respiratory activity in pea cotyledons during seed development. *Journal of Experimental Botany* 34, 1026–1036.

Konno, S. (1979). Changes in chemical composition of soybean seeds during ripening. *Japanese Agricultural Research Quarterly* 13, 186–194.

Kowitt, B. (2014). Just Mayo hits the big time. *Fortune*, 30 July. Available at: http://www.fortune.com/2014/07/30/more-than-just-mayo (accessed 18 June 2016).

Krenzer, E.G. and Moss, D.N. (1975). Carbon dioxide enrichment effects upon yield and yield components in wheat. *Crop Science* 15, 71–74.

Kropft, M.J., Haverkort, A.J., Aggarwal, P.K. and Kooman, P.L. (1995). Using system approaches to design and evaluate ideotypes for specific environments. In: Bouman, J., Kuyvenhoven, A., Bouman, B., Luyten, J. and Zandstra, H. (eds) *Eco–regional Approaches for Sustainable Land Use and Food Production*. Kluwar Academic Publishers, Dordrecht, The Netherlands, pp. 417–435.

Kupferschmidt, K. (2015). Buzz food. *Science* 350, 267–269.

Lake, L. and Sadras, V.O. (2014). The critical period for yield determination in chickpea (*Cicer arietinum* L.). *Field Crops Research* 168, 1–7.

Langer, R.H.M. and Hill, G.D. (1991). *Agricultural Plants*, 2nd edn. Cambridge University Press. Cambridge, UK.

Larson, E.M., Hesketh, J.D., Woodley, J.T. and Peters, D.B. (1981). Seasonal variation in apparent photosynthesis among plant stands of different soybean cultivars. *Photosynthesis Research* 2, 3–20.

Lawn, R.J. (1989). Agronomic and physiological constraints to the productivity of tropical grain legumes and prospects for improvements. *Experimental Agriculture* 25, 509–528.

Lawn, R.J. and Brun, W.A. (1974). Symbiotic nitrogen fixation in soybeans. I. Effect of photosynthetic source–sink manipulation *Crop Science* 14, 11–16.

Lawn, R.J. and Imrie, B.C. (1991). Crop improvement for tropical and subtropical Australia: Designing plants for difficult climates. *Field Crops Research* 26, 113–139.

Layzell, D.B. and LaRue, T.A. (1982). Modeling C and N transport to developing soybean fruits. *Plant Physiology* 70, 1290–1298.

Lee, H.J., McKee, G.W. and Knievel, D.P. (1979). Determination of physiological maturity in oat. *Agronomy Journal* 71, 931–935.

Lemcoff, J.H. and Loomis, R.S. (1986). Nitrogen influences on yield determination in maize. *Crop Science* 26, 1017–1022.

Leon, L. and Geisler, G. (1994). Variation in rate and duration of growth among spring barley cultivars. *Plant Breeding* 112, 199–208.

Lersten, N.R. (1987). Morphology and anatomy of the wheat plant. In: Heyne E.G. (ed.) *Wheat and Wheat Improvement*. 2nd edn. Crop Science Society of America. Madison, Wisconsin USA, pp. 33–75.

Lhuillier-Soundele, N.G., Munier-Jolian, N.G. and Ney, B. (1999). Dependence of seed nitrogen concentration on plant nitrogen availability during seed filling in pea. *European Journal of Agronomy* 11, 157–166.

Lim, P.O., Kim, H.J. and Nam, H.G. (2007). Leaf senescence. *Annual Review of Plant Biology* 58, 115–136.

Lin, M. and Huybers, P. (2012). Reckoning wheat yield trends. *Environmental Research Letters* 112, 1–7.

Lindstrom, L.I., Pellegrini, C.N., Aguirrezaba, L.A.N. and Hernandez, L.F. (2006). Growth and development of sunflower fruits under shade during pre- and early post-anthesis period. *Field Crops Research* 96, 151–159.

Liu, F., Jensen, C.R. and Anderson, M.A. (2004). Drought stress effect on carbohydrate concentration in soybean leaves and pods during early reproductive development: Implications in altering pod set. *Field Crops Research* 86, 1–13.

Lockhart, J.A. (1965). An analysis of irreversible plant cell elongation. *Journal of Theoretical Biology* 8, 264–275.

Ludlow, M.M. and Muchow, R.C. (1990). A critical evaluation of traits for improving crop yields in water–limited environments. *Advances in Agronomy* 43, 107–153.

Lush, W.M. and Evans, L.T. (1981). The domestication and improvement of cowpeas (*Vigna unguiculata* L. Walp). *Euphytica* 30, 579–587.

Ma, Y.-Z., MacKown, C.T. and Van Sandford, D.A. (1995). Kernel mass and assimilate accumulation of wheat: Cultivar response to 50% spikelet removal at anthesis. *Field Crops Research*. 42, 93–99.

Ma, M., Wang, Q., Li, Z., Cheng, H., Li, Z., Liu, X., Song, W., Appels, R. and Zhao, H. (2015). Expression of TaCyP78A3 , a gene incoding cytochrome P450 CYP78A3 protein in wheat (*Triticum aestivum* L.) affects seed size. *The Plant Journal* 83, 312–325.

Malik, T.A., Ali, Y., Saleem, M. and Tahir, G.R. (1989). Yield stability of induced-early-maturing mutants of mungbean (*Vigna radiata*) and their use in multiple cropping systems. *Field Crops Research* 20, 251–259.

Malthus, T.R. (1993 [1798]). *An Essay of the Principle of Population.* Oxford World's Classics. Oxford University Press, Oxford, UK.

Mangelsdorf, P.C., MacNeish, R.S. and Galinal, W.G. (1967). *Prehistoric Wild and Cultivated Maize.* University of Texas, Austin, Texas, USA.

Mann, J.D. and Jaworski, E.G. (1970). Comparison of stress which may limit soybean yields. *Crop Science* 10, 620–624.

Marshall, D.R. (1991). Alternative approaches and perspectives in breeding for higher yield. *Field Crops Research* 26, 171–190.

Marte, P., Quilot-Turion, B., Luquet, D., Ould-Sidi Memmah, M.-M., Chenu, K. and Debaeke, P. (2015). Model-assisted phenotyping and ideotype design. In: Sadras, V.O. and Calderine, D.F. (eds) *Crop Physiology. Applications for Genetic Improvement and Agronomy,* 2nd edn. Academic Press, London, UK, pp. 349–373.

May, L. and Van Sanford, D.A. (1992). Selection for early heading and correlated response in maturity of soft red winter wheat. *Crop Science* 32, 47–51.

McBlain, B.A. and Hume, D.J. (1980). Physiological studies of higher yield in new early-maturing soybean cultivars. *Canadian Journal of Plant Science* 60, 1315–1326.

McBlain, B.A. and Hume, D.J. (1981). Reproductive abortion, yield components, and nitrogen content in three early soybean cultivars. *Canadian Journal of Plant Science* 61, 499–505.

McDonald, M.B. and Copeland, L.O. (1997). *Seed Production. Principles and Practices,* 3rd edn. Chapman and Hall, New York, USA.

McGarrahan, J.P. and Dale, R.F. (1984). A trend toward a longer grain-filling period for corn: A case study in Indiana. *Agronomy Journal* 76, 518–522.

Meckel, L., Egli, D.B., Phillips, R.E., Radcliffe, D. and Leggett, J.E. (1984). Effect of moisture stress on seed growth in soybeans. *Agronomy Journal* 75, 1027–1031.

Metz, G.L., Green, D.E. and Shibles, R.M. (1984). Relationship between soybean yield in narrow rows and leaflet, canopy and developmental characters. *Crop Science* 24, 457–462.

Metz, G.L., Green, D.E. and Shibles, R.M. (1985). Reproductive duration and date of maturity in populations of three wide soybean crosses. *Crop Science* 25, 171–176.

Metzger, D.D., Czaplewski, S.J. and Rasmusson, D.C. (1984). Grain-filling duration and yield in spring barley. *Crop Science* 24, 1101–1105.

Miceli, F., Crafts-Brandner, S.J. and Egli, D.B. (1995). Physical restriction of pod growth alters development of soybean plants. *Crop Science* 35, 1080–1085.

Miflin, B. (2000). Crop improvement in the 21st century. *Journal of Experimental Botany* 51, 1–8.

Miller, F.R. and Kebede, Y. (1984). Genetic contributions to yield gains in sorghum, 1950 to 1980. In: Fehr, W.R. (ed.) *Genetic Contributions to Yield Gains of Five Major Crop Plants.* Spec. Pub. no. 7. Crop Science Society of America, Madison, Wisconsin, USA, pp. 1–14.

Millet, E. (1986). Relationships between grain weight and the size of floret cavity in the wheat spike. *Annals of Botany* 58, 417–423.

Millet, E. and Pinthus, M.J. (1984). Effects of removing floral organs, light penetration and physical constraint on the development of wheat grains. *Annals of Botany* 53, 261–270.

Miralles, D.J. and Slafer, G.A. (1995). Individual grain weight responses to genetic reduction in culm length in wheat as affected by source–sink manipulations. *Field Crops Research* 43, 55–66.

Mitchell, J.H., Fukai, S. and Cooper, M. (1996). Influence of phenology on grain yield variation among barley cultivars grown under terminal drought. *Australian Journal of Agricultural Research* 47, 757–774.

Moles, A.T., Ackerlt, D.C., Webb, C.O., Tweddle, J.C., Dickie, J.B. and Westoby, M. (2005). A brief history of seed size. *Science* 307, 576–580.

Mondal, M.H., Brun, W.A. and Brenner, M.L. (1978). Effects of sink removal on photosynthesis and senescence in leaves of soybean (*Glycine max* L.) plants. *Plant Physiology* 61, 394–397.

Monk, R., Franks, C. and Dahlberg, J. (2014). Sorghum. In: Smith, S., Diers, B.W., Specht, J.E. and Carver, B. (eds) *Yield Gains in Major U.S. Field Crops.* ASA-CSSA-SSSA, Madison, Wisconsin, USA, pp. 293–310.

Monteith, J.L. (1978). Reassessment of maximum growth rates for C3 and C4 crops. *Experimental Agriculture* 14, 1–5.

Morandi, E.N., Casano, L.M. and Reggiardo, L.M. (1988). Post-flowering photoperiodic effect on reproductive efficiency and seed growth in soybean. *Field Crops Research* 18, 227–241.

Motto, M. and Moll, R.H. (1983). Prolificacy in maize: A review. *Maydica* 28, 53–76.

Motzo, R., Giunta, F. and Pruneddy, G. (2010). The response of rate and duration of grain filling to long-term selection for yield in Italian durum wheats. *Crop and Pasture Science* 61, 162–169.

Mou, B. and Kronstad, W.E. (1994). Duration and rate of grain filling in selected winter wheat populations: I. Inheritance. *Crop Science* 34, 833–837.

Muchow, R.C., Sinclair, T.R. and Bennett, J.M. (1990). Temperature and solar radiation effects on potential maize yield across locations. *Agronomy Journal* 82, 338–343.

Mulatu, E. and Belete, K. (2001). Participatory varietal selections in low land sorghum in Ethiopia: Impact on adoption and genetic diversity. *Experimental Agriculture* 37, 211–229.

Munier-Jolain, N.G. and Ney, B. (1998). Seed growth rate in grain legumes II. Seed growth rate depends on cotyledon cell number. *Journal of Experimental Botany* 49, 1971–1976.

Munier-Jolain, N.G., Ney, B. and Duthion, C. (1993). Sequential development of flowers and seeds on the main stem of an indeterminate soybean. *Crop Science* 33, 768–771.

Munier-Jolain, N.G., Ney, B. and Duthion, C. (1996). Termination of seed growth in relation to nitrogen content of vegetative parts in soybean plants. *European Journal of Agronomy* 5, 219–225.

Munier-Jolain, N.G., Munier-Jolain, N.M., Roche, R. and Ney, B. (1998). Seed growth in grain legumes I. Effect of photoassimilates availability on seed growth rate. *Journal of Experimental Botany* 49, 1963–1969.

Murata, Y. (1969). Physiological responses to nitrogen in plants. In: Eastin, J.D., Haskins, F.A., Sullivan, C.Y. and Van Bavel, C.H.M. (eds) *Physiological Aspects of Crop Yield.* American Society of Agronomy – Crop Science Society of America. Madison, Wisconsin USA, pp. 235–259.

Murata, Y. (1981). Dependence of potential productivity and efficiency for solar energy utilization on leaf photosynthetic capacity in crop species. *Japanese Journal of Crop Science* 50, 223–232.

Murata, Y. and Matsushima, S. (1975). Rice. In: Evans, L.T. (ed.) *Crop Physiology.* Cambridge University Press. Cambridge UK, pp. 73–99.

NASS (2016). National Agricultural Statistics Service. Available at: https://www.nass.usda.gov (accessed 18 June 2016).

NeSmith, D.S. and Ritchie, J.T. (1992). Maize (*Zea mays* L.) response to a severe soil water-deficit during grain-filling *Field Crops Research* 29, 23–35.

Nestle, M. (2013). *Food Politics. How the Food Industry Influences Nutrition and Health,* 10th edn. University of California Press, Berkeley, California, USA.

Ney, B., Duthion, C. and Fontaine, E. (1993). Timing of reproductive abortions in relation to cell division, water content, and growth of pea seeds. *Crop Science* 33, 267–270.

Ney, B., Duthion, C. and Turc, O. (1994). Phenological response of pea to water stress during reproductive development. *Crop Science* 34, 141–146.

Neyshabouri, M.R. and Hatfield, J.L. (1986). Soil water deficit effects on semi-determinate and indeterminate soybean growth and yield. *Field Crops Research* 15, 73–84.

Nico, M., Miralles, D.J. and Kantolic, A.G. (2015). Post-flowering photoperiod and radiation interaction in soybean yield determination: Direct and indirect effects. *Field Crops Research* 176, 45–55.

Nicolas, M.E., Gleadow, R.M. and Dalling, M.J. (1985). Effect of post-anthesis drought on cell division and starch accumulation in developing wheat grains. *Annals of Botany* 55, 433–444.

NOAA (2016). Climate Data Online: Dataset Discovery. National Oceanic and Atmospheric Administration. Available at: http://ncdc.noaa.gov/cdo-web/datasets#Normal_Ann (accessed 27 August 2016).

Normile, D. (2016). Nature from nuture. *Science* 351, 908–910.

Obendorf, R.L., Ashworth, E.N. and Rytko, G.T. (1980). Influence of seed maturation on germinability in soybean. *Crop Science* 20, 483–486.

O'Grady, C. (2009). *Famine. A Short History.* Princeton University Press. Princeton, New Jersey USA.

Ohmura, T. and Howell, R.W. (1962). Respiration of developing and germinating soybean seeds. *Physiologia Plantarum* 15, 341–350.

Oparka, K.J. and Gates, P. (1981). Transport of assimilates in the developing caryopsis of rice (*Oryza sativa* L.). Ultrastructure of the pericarp vascular bundle and its connections with the aleurone layer. *Planta* 151, 561–573.

Ortiz-Monasterio, J.I., Dhillon, S.S. and Fischer, R.A. (1994). Date of sowing effects on grain yield and yield components of irrigated spring wheat cultivars and relationships with radiation and temperature in Ludhjiana, India. *Field Crops Research* 37, 169–184.

Oscarson, P. (2000). The strategy of the wheat plant in acclimating growth and grain production to nitrogen availability. *Journal of Experimental Botany* 51, 1921–1929.

Otegui, M.E. (1995). Prolificacy and grain yield components in modern Argentinean maize hybrids. *Maydica* 40, 371–376.

Otegui, M.E. and Andrade, F.H. (2000). New relationships between light interception, ear growth, and kernel set in maize. In: Westgate, M.E. and Boote, K.J. (eds) *Physiology and Modeling Kernel Set in Maize*. CSSA, Madison, Wisconsin, USA, pp. 89–102.

Lim, P.O., Kim, H.J. and Nam, H.G. (2007). Leaf senescence. *Annual Review of Plant Biology* 58, 115–136.

Paddock, W. and Paddock, P. (1967). *Famine in 1975*. Little, Brown & Co., Boston, Massachusetts, USA.

Pandy, R.K., Herrera, W.A.T. and Pendleton, J.W. (1984a). Drought response of grain legumes under irrigation gradient. I. Yield and yield components. *Agronomy Journal* 76, 549–553.

Pandy, R.K., Herrera, W.A.T. and Pendleton, J.W. (1984b). Drought response of grain legumes under irrigation gradient. II. Plant water status and canopy temperature. *Agronomy Journal* 78, 553–560.

Pask, A.J.D. and Reynolds, M.P. (2013). Breeding for yield potential has increased deep soil water extraction capacity in irrigated wheat. *Crop Science* 53, 2090–2104.

Passioura, J.B. (1996). Simulation models: Science, snake oil, education or engineering? *Agronomy Journal* 88, 690–694.

Patrick, J.W. and Offler, C.E. (2001). Compartmentation of transport and transfer events in developing seeds. *Journal of Experimental Botany* 52, 551–564.

Pearson, C.J., E.M. Larson, Hesketh, J.D. and Peters, D.B. (1984). Development and source–sink effects on single leaf and canopy carbon dioxide exchange in maize. *Field Crops Research* 9, 391–402.

Peaslee, D.E., Ragland, J.L. and Duncan, W.G. (1971). Grain filling period of corn as influenced by phosphorus, potassium, and the time of planting. *Agronomy Journal* 63, 561–563.

Peltonen-Sainio, P. (1991). Productive oat ideotype for northern growing conditions. *Euphytica* 54, 27–32.

Peltonen-Sainio, P. (1993). Contribution of enhanced growth rate and associated physiological changes to yield formation of oats. *Field Crops Research* 33, 269–281.

Peng, S. and Khush, G.S. (2003). Four decades of breeding for varietal improvement of irrigated lowland rice in the International Rice Research Institute. *Plant Science* 6, 157–164.

Peng, S., Laza, R.C., Visperas, R.M., Sancco, A.L., Cassman, K.G. and Krush, G.S. (2000). Grain yield of rice cultivars and lines developed in the Philippines since 1966. *Crop Science* 40, 307–314.

Penning de Vries, F.W.T., Brunsting, A.H.M. and van Laar, H.H. (1974). Products, requirements and efficiency of biosynthesis: A quantitative approach. *Journal of Theoretical Biology* 45, 339–377.

Penrose, L.D., Walsh, J. and Clarck, K. (1998). Characters contributing to high yield in Currawong, an Australian winter wheat. *Australian Journal of Agricultural Research* 49, 853–866.

Pepler, S., Gooding, M.J., Ford, K.E. and Ellis, R.H. (2005). A temporal limit to the association between flag leaf life extension by fungicides and wheat yields. *European Journal of Agronomy* 22, 363–373.

Peters, D.B., Pendleton, J.W., Hageman, R.H. and Brown, C.M. (1971). Effect of night air temperatures on grain yield of corn, wheat, and soybeans. *Agronomy Journal* 63, 809.

Pfeiffer, T.W. and Egli, D.B. (1988). Heritability of seed-filling period estimates in soybean. *Crop Science* 28, 921–925.

Pfeiffer, T.W., Suryati, D. and Egli, D.B. (1991). Soybean plant introductions selected for seed filling period or yield performance as parents. *Crop Science* 31, 1418–1421.

Phakamas, N., Patanothai A., Jogloy, S., Pannanagpetch, K. and Hoogenboom, G. (2008). Physiologic determinants for pod yield of peanut lines. *Crop Science* 48, 2351–2360.

Pieta-Filho, C. and Ellis, R.H. (1991). The development of seed quality in spring barley in four environments. I. Germination and longevity. *Seed Science Research* 1, 163–177.

Plucknett, D.L., Everson, J.P. and Sanford, W.G. (1970). Ratoon cropping. *Advances in Agronomy* 22, 285–330.

Poggio, S.L., Satorre, E.H., Dethiou, S. and Gonzalo, G.M. (2005). Pod and seed numbers as a function of photothermal quotient during the seed set period of field pea (*Pisum sativum*). *European Journal of Agronomy* 22, 55–69.

Pokojska, H. and Grzelak, K. (1996). Influence of seed maturity on germination, vigor and protein and tannin contents in faba bean (*Vicia faba* L. var. minor). *Plant Breeding and Seed Science* 40, 11–20.

Poneleit, C.G. and Egli, D.B. (1979). Kernel growth rate and duration in maize as affected by plant density and genotype. *Crop Science* 19, 385–388.

Porter, P.M., Leuer, J.G., Huggins, D.R., Oplinger, E.S. and Crookston, R.K. (1998). Assessing spatial and temporal variability of corn and soybean yields. *Journal of Production Agriculture* 11, 369–365.

Prine, G.M. (1971). A critical period for ear development in maize. *Crop Science* 11, 782–786.

Pszczola, D.E. (2012). Seeds of success. *Food Technology* 66, 45–55.

Puckridge, D.W. and Donald, C.M. (1967). Competition among wheat plants sown at a wide range of densities. *Australian Journal of Agriculture Research* 18, 193–211.

Purcell, L.C. (2008). Resources needed for record–breaking soybean yields. In: Pringnitz B.A. (ed.) *20th Annual Integrated Crop Management Conference*. Iowa State University, Ames, Iowa, pp. 17–20.

Quattar, S., Jones, R.J. and Crookston, R.K. (1987). Effect of water deficit during grain filling on the pattern of maize kernel growth and development. *Crop Science* 27, 726–736.

Rainbird, R.M., Thorne ,J.H. and Hardy, R.W.F. (1984). Role of amides, amino acids, and ureides in the nutrition of developing soybean seeds. *Plant Physiology* 74, 329–334.

Rajala, A. and Peltonen-Sainio, P. (2004). Intra-plant variation in progress of cell division in developing oat grains: A preliminary study. *Agriculture and Food Science* 13, 163–169.

Rajcan, I. and Tollenaar, M. (1999). Source: sink ratio and leaf senescence in maize. I. Dry matter accumulation and partitioning during grain filling. *Field Crops Research* 60, 245–253.

Rajewski, J.F. and Francis, C.A. (1991). Defoliation effects on grain fill, stalk rot and lodging of grain sorghum. *Crop Science* 31, 353–359.

Ramseur E.L., Wallace S.U. and Quisenberry V.L. (1985). Growth of 'Braxton' soybeans as influenced by irrigation and intrarow spacing. *Agronomy Journal* 77, 163–168.

Rasmusson, D.C., McLean, I. and Tew, T.L. (1979). Vegetative and grain-filling periods of growth in barley. *Crop Science* 19, 5–9.

Rawson, H.M. (1996). The developmental stage during which boron limitation causes sterility in wheat genotypes and the recovery of fertility. *Australian Journal of Plant Physiology* 23, 709–717.

Rawson, H.M. and Bagga, A.K. (1979). Influence of temperature between floral initiation and flag leaf emergence on grain number in wheat. *Australian Journal of Plant Physiology* 6, 391–400.

Rawson, H.M. and Evans, L.T. (1970). The pattern of grain growth within the ear of wheat. *Australian Journal of Biological Science* 23, 753–764.

Ray, D.K., Mueller, N.D., West, P.C. and Foley, J.A. (2013). Yield trends are insufficient to double crop production by 2050. *PLoS ONE* 8(6), e66428. DOI: 10.1371/journal.pone.0066428.

Reddy, V.M. and Daynard, T.B. (1983). Endosperm characteristics associated with rate of grain filling and kernel size in corn. *Maydica* 28, 339–355.

Rench, W.E. and Shaw, R.H. (1971). Black layer development in corn. *Agronomy Journal* 63, 303–309.

Reynolds, M., Foulkes, M.J., Slafer, G.A., Parry, M.A.J., Snape, J.W. and Angus, W.J. (2009). Raising yield potential in wheat. *Journal of Experimental Botany* 60, 1899–1918.

Richards, R.A. (2006). Physiological traits used in the breeding of new cultivars for water-scarce environments. *Agricultural Water Management* 80, 197–211.

Ritchie, J.T. and Wei, J. (2000). Models of kernel number in maize. In: Westgate, M.E. and Boote, K.J. (eds) *Physiology and Modeling Kernel Set in Maize*. Crop Science Society of America, Madison, Wisconsin, USA, pp. 75–88.

Roberts E. (1847). On the management of wheat. *Journal of the Royal Agriculture Society* 60–77.

Robinson, R.G. (1983). Maturation of sunflower and sector sampling of heads to monitor maturation. *Field Crops Research* 7, 31–39.

Rochat, C. and Boutin, J.P. (1991). Metabolism of phloem-borne amino acids in maternal tissues of fruit of nodulated or nitrate-fed pea plants (*Pisum sativum* L.). *Journal of Experimental Botany* 42, 207–214.

Rodriquez-Sotres, R. and Black, M. (1994). Osmotic potential and abscisic acid regulate triacyglycerol synthesis in developing wheat embryos. *Planta* 192, 9–15.

Rondanini, D., Savin, R. and Hall, A.J. (2007). Estimation of physiological maturity in sunflower as a function of fruit water concentration. *European Journal of Agronomy* 26, 295–309.

Rotundo, J.L. and Westgate, M.E. (2010). Rate and duration of seed component accumulation in water-stressed soybean. *Crop Science* 50, 676–684.

Ruan, Y.-L., Patrick, J.W., Bouzyan, M., Osorio, S. and Fernie, A. (2012). Molecular regulation of seed and fruit set. *Trends in Plant Science* 17, 656–665.

Ruiz, R.A. and Maddonni, G.A. (2006). Sunflower seed weight and oil concentration under different post-flowering source–sink ratios. *Crop Science* 46, 671–680.

Russell, W.A. (1991). Genetic improvement in maize yields. *Advances in Agronomy* 46, 245–298.

Saab, I.N. and Obendorf, R.L. (1989). Soybean seed water relations during *in situ* and *in vitro* growth and maturation. *Plant Physiology* 89, 610–616.

Sadras, V.O. and Conner, D.J. (1991). Physiological basis of the response of harvest index to the fraction of water transpired after anthesis: A simple model to estimate harvest index for determinate species. *Field Crops Research* 26, 227–240.

Saini, H.S. and Westgate, M.E. (2000). Reproductive development in grain crops during drought. *Advances in Agronomy* 68, 59–96.

Sala, R.G., Westgate, M.E. and Andrade, F.H. (2007a). Source/sink ratio and the relationship between maximum water content, maximum volume and final dry weight of maize kernels. *Field Crops Research* 101, 19–25.

Sala, R.G., Andrade, F.H. and Westgate, M.E. (2007b). Maize kernel moisture at physiological maturity as affected by source–sink relationships during grain filling. *Crop Science* 47, 711–714.

Salado-Navarro, L.R., Sinclair, T.R. and Hinson, K. (1985). Comparisons among effective filling period, reproductive period duration, and R5–R7 in determinate soybeans. *Crop Science* 25, 1050–1054.

Sambo, E.Y., Moorby, J. and Milthrope, F.L. (1977). Photosynthesis and respiration of developing soybean pods. *Australian Journal of Plant Physiology* 4, 713–721.

Sandfaer, J. and Haahr, V. (1975). Barley stripe mosaic virus and the yield of old and new barley varieties. *Zeitschrift für Pflanzenzüchtung* 74, 211–222.

Satake, T. and Yoshida, S. (1978). High temperature induced sterility in indica rices at flowering. *Japanese Journal of Crop Science* 47, 6–17.

Schapaugh, W.T. and Wilcox, J.R. (1980). Relationships between harvest indices and other plant characteristics in soybean. *Crop Science* 20, 529–533.

Schmidt, J.W. (1984). Genetic contributions to yield gains in wheat. In: Fehr, W.R. (ed.) *Genetic Contributions to Yield Gains of Five Major Crop Plants.* Spec. Pub. no. 7. Crop Science Society of America, Madison, Wisconsin, USA, pp. 89–101.

Schneiter, A.A. and Miller, J.F. (1981). Description of sunflower growth stages. *Agronomy Journal* 21, 901–963.

Schnyder, H. and Baum, U. (1992). Growth of the grain of wheat (*Triticum aestivum* L.). The relationship between water content and dry matter accumulation *European Journal of Agronomy* 1, 51–57.

Schoper, J.B., Johnson, R.R. and Lambert, R.J. (1982). Maize yield response to increased assimilate supply. *Crop Science* 22, 1184–1189.

Schou, J.B., Jeffers, D.L. and Streeter, J.G. (1978). Effects of reflectors, black boards, or shades applied at different stages of plant development on yield of soybeans. *Crop Science* 18, 29–34.

Schussler, J.R. and Brenner, M.L. (1989). Regulation of soybean seed growth and development by ABA. In: Pascale, A.J. (ed.) *Proceedings of Fourth World Soybean Research Conference.* Asociación Argentina de la Soja, Buenos Aires, Argentina, pp. 262–271.

Schussler, J.R. and Westgate, M.E. (1994). Increasing assimilate reserves does not prevent kernel abortion at low water potential in maize. *Crop Science* 34, 1596–1576.

Schweitzer, L.E. and Harper, J.E. (1985). Effect of hastened flowering on seed yield and dry matter partitioning in diverse soybean genotypes. *Crop Science* 25, 995–998.

Scott, W.R., Appley, M., Fellowes, G. and Kirby, E.J.M. (1983). Effect of genotype and position in the ear on carpel and grain growth and mature grain weight in spring barley. *Journal of Agricultural Science (Cambridge)* 100, 383–391.

Secor, J., Shibles, R.M. and Stewart, C.R. (1983). Metabolic changes in senescing soybean leaves of similar plant ontogeny. *Crop Science* 23, 106–109.

Sedgley, R.H. (1991). An appraisal of the Donald ideotype after 21 years. *Field Crops Research* 26, 93–112.

Seufert, V., Ramankutty, N. and Foley, J.A. (2012). Comparing yields of organic and conventional agriculture. *Nature* 485, 229–234.

Sexton, P.J., White, J.W. and Boote, K.J. (1994). Yield determining processes in relation to cultivar seed size of common bean. *Crop Science* 34, 84–91.

Shackel, K.A. and Turner, N.C. (2000). Seed coat cell turgor in chickpea is independent of changes in plant and pod water potential. *Journal of Experimental Botany* 51, 895–900.

Sharma, R.C. (1994). Early generation selection for grain filling period in wheat. *Crop Science* 34, 945–948.

Shaw, R.H. and Laing, D.R. (1966). Moisture stress and plant response. In: Pierre, W.H., Kirkham, D., Pesek, J. and Shaw, R. (eds) *Plant Environment and Efficient Water Use.* American Society of Agronomy, Madison, Wisconsin, USA, pp. 73–94.

Shaw, R.H. and Loomis, W.E. (1950). Basis for the prediction of corn yields. *Plant Physiology* 25, 225–244.

Shearman, V.J., Sylvester-Bradley, R., Scott, R.K. and Foulkes, M.J. (2005). Physiological processes associated with wheat yield progress in the U.K. *Crop Science* 45, 175–185.

Sheldrake, A.R. (1979). A hydrodynamical model of pod-set in pigeon pea (*Cajanus cajan*). *Indian Journal of Plant Physiology* 22, 137–143.

Shibles, R.M., Anderson, I.C. and Gibson, A.H. (1975). Soybean. In: Evans, L.T. (ed.) *Crop Physiology: Some Case Histories*. Cambridge University Press, Cambridge, UK, pp. 151–190.

Shibles, R.M. and Weber, C.R. (1965). Leaf area, solar radiation interception and dry matter production by soybeans. *Crop Science* 5, 575–577.

Shimelis, H., Mashela, P.W. and Hugo, A. (2008). Performance of Vernonia as an alternative industrial oil crop in Limpopo province of South Africa. *Crop Science* 48, 236–242.

Shiraiwa, T. and Hashikawa, U. (1995). Accumulation and partitioning of nitrogen during seed filling in old and modern soybean cultivars in relation to seed production. *Japanese Journal of Crop Science* 64, 754–759.

Siemer, E.G. (1964). Major developmental events in maize – their timing, correlation, and mature plant expression. PhD thesis. University of Illinois, Urbana, Illinois, USA.

Simmons, S.R., Crookston, R.K. and Kurle, J.E. (1982). Growth of spring wheat kernels as influenced by reduced kernel number per spike and defoliation. *Crop Science* 22, 983–988.

Sinclair, T.R. (1980). Leaf CER from post-flowering to senescence of field-grown soybean cultivars. *Crop Science* 20, 196–200.

Sinclair, T.R. (1986). Water and nitrogen limitations in soybean production. I. Model development. *Field Crops Research* 15, 125–141.

Sinclair, T.R. and Bai, Q. (1997). Analysis of high wheat in Northwest China. *Agricultural Systems* 53, 373–385.

Sinclair, T.R. and de Wit, C.T. (1975). Comparative analysis of photosynthate and nitrogen requirements in the production of seeds by various crops. *Science* 18, 565–567.

Sinclair, T.R. and Jamieson, P.D. (2008). Grain number, wheat yield, and bottling beer. An analysis. *Field Crops Research* 98, 60–67.

Sinclair, T.R. and Purcell, L.C. (2005). Is a physiological perspective relevant in a 'genocentric' age? *Journal of Experimental Botany* 56, 2777–2782.

Sinclair, T.R. and Sinclair, C.J. (2010). *Bread, Beer and the Seeds of Change. Agriculture's Imprint on World History*. CAB International, Wallingford, Oxfordshire, UK.

Sinclair, T.R., Purcell, L.C. and Sneller, C.H. (2004). Crop transformation and challenge to increase yield potential. *Trends in Plant Science* 9, 70–75.

Singh, B.K. and Jenner, C.F. (1982). Association between concentrations of organic nutrients in the grain, endosperm cell number and grain dry weight within the ear of wheat. *Australian Journal of Plant Physiology* 9, 83–95.

Singh, B.K. and Jenner, C.F. (1984). Factors controlling endosperm cell number and grain dry weight in wheat: Effects of shading on intact plants and variation in nutritional supply to detached, cultured ears. *Australian Journal of Plant Physiology* 11, 151–163.

Singletary, G.W. and Below, F.E. (1989). Growth and composition of maize kernels cultured *in vitro* with varying supplies of carbon and nitrogen. *Plant Physiology* 89, 341–346.

Slafer, G.A. and Savin, R. (1994). Source–sink relationships and grain mass at different positions within the spike in wheat. *Field Crops Research* 37, 39–34.

Slafer, G.A., Satorre, E.H. and Andrade, F.H. (1994a). Increases in grain yield of bread wheat from breeding and associated physiological changes. In: Slafer G.A. (ed.) *Genetic Improvement of Field Crops*, Marcell Dekker Inc., New York, USA, pp. 1–68.

Slafer, G.A., Calderni, D.F., Miralles, D.J. and Dreccer, M.F. (1994b). Preanthesis shading effects on the number of grains of three bread wheat cultivars of different potential number of grains. *Field Crops Research* 36, 31–39.

Slafer, G.A., Kantolic, A.G., Appendino, M.L., Miralles, D.J. and Savin, R. (2009). Crop development; Genetic control, environmental modulation and relevance for genetic improvement of crop yield. In: Sadras, V.O. and Calderine, D.F. (eds) *Crop Physiology. Applications for Genetic Improvement and Agronomy.* Academic Press, New York, USA, pp. 277–308.

Slafer, G.A., Savin, R. and Sadras, V.O. (2014). Coarse and fine regulation of wheat yield components in response to genotype and environment. *Field Crops Research* 157, 71–83.

Smith, D.H. and Donnelly, K.J. (1991). Visual markers of maximum grain dry weight accumulation in winter wheat. *Crop Science* 31, 419–422.

Smith, J.R. and Nelson, R.L. (1986a). Relationship between seed filling period and yield among soybean breeding lines. *Crop Science* 26, 469–472.

Smith, J.R. and Nelson, R.L. (1986b). Selection for seed-filling period in soybeans. *Crop Science* 26, 466–469.

Smith, J.R. and Nelson, R.L. (1987). Predicting yield from early generation estimates of reproductive growth periods in soybean. *Crop Science* 27, 471–474.

Smith, S., Cooper, M., Gogerty, J., Loffler, C., Borcherding, D. and Wright, K. (2014). Maize. In: Smith, S., Diers, B., Specht, J. and Carver, B. (eds) *Yield Gains in Major U.S. Field Crops.* ASA-CSSA-SSSA, Madison, Wisconsin, USA, pp. 125–171.

Sofield, I., Evans, L.T., Cook, M.G. and Wardlaw, I.F. (1977). Factors influencing the rate and duration of grain filling in wheat. *Australian Journal of Plant Physiology* 4, 785–797.

Sorrells, M.E. and Meyers, O.J. (1982). Duration of developmental stages of 10 milo maturity genotypes. *Crop Science* 22, 310–314.

Spaeth, S.C. and Sinclair, T.R. (1984). Soybean seed growth. I. Timing of growth of individual seeds. *Agronomy Journal* 76, 123–127.

Spears, J.F., TeKrony, D.M. and Egli, D.B. (1997). Temperature during seed filling and soybean seed germination and vigour. *Seed Science and Technology* 25, 233–244.

Specht, J.E., Diers, B.W., Nelson, R.L., de Toledo, J.A., Torrion, J.A. and Grassini, P. (2014). Soybean. In: Smith, S., Diers, B., Specht, J. and Carver, B. (eds) *Yield Gains in Major U.S. Field Crops.* ASA-CSSA-SSSA, Madison, Wisconsin, USA, pp. 311–355.

Stanhill, G. (1976). Trends and deviations in the yield of the English wheat crop during the last 750 years. *Agro-Ecosystems* 3, 1–10.

Stapper, M. and Arkin, G.F. (1980). *CORNF: A Dynamic Growth and Development Model for Maize (Zea mays L.).* Program and Model. DOC 80-2 (ed). Texas Agriculture Experiment Station, Temple, Texas, USA.

Steer, B.T., Hocking, P.J. and Low, A. (1988). Dry matter, minerals and carbohydrates in the capitulum of sunflower (*Helianthus annus*). Effects of competition between seeds, and defoliation. *Field Crops Research* 18, 71–85.

Sung, F.J.M. and Chen, J.J. (1990). Cotyledon cells and seed growth relationships in CO_2 enriched peanut. *Peanut Science* 17, 4–6.

Swank, J.C., Below, F.E., Lambert, R.J. and Hageman, R H. (1982). Interaction of carbon and nitrogen metabolism in the productivity of maize. *Plant Physiology* 70, 1185–1190.

Swank, J.C., Egli, D.B. and Pfeiffer, T.W. (1987). Seed growth characteristics of soybean genotypes differing in duration of seed fill. *Crop Science* 27, 85–89.

Talbert, L.E., Lanning, S.P., Murphy, R.L. and Martin, J.M. (2001). Grain fill duration in twelve hard red spring wheat crosses: Genetic variation and association with other traits. *Crop Science* 41, 1390–1395.

Tanaka, A. (1980). Source and sink relationships in crop production. In: Center F.F.T. (ed.) Technical Bulletin No. 52, Taipei City, Taiwan.

Tang, T., Xie, H., Wang, Y., Liu, B. and Laing, J. (2009). The effect of sucrose and abscisic acid interaction on sucrose synthase and its relationship to grain filling of rice (*Oryza sativa* L.). *Journal of Experimental Botany* 60, 2641–2652.

Tanner, J.W. and Ahmed, S. (1974). Growth analysis of soybeans treated with TIBA. *Crop Science* 14, 371–374.

Tashiro, T. and Wardlaw, I.F. (1989). A comparison of the effect of high temperature on grain development in wheat and rice. *Annals of Botany* 64, 59–65.

Taylor, A.M., Amiro, B.D. and Fraser, T.J. (2013). Net CO_2 exchange and carbon budgets of a three-year crop rotation following conversion of perennial lands to annual cropping in Manitoba, Canada. *Agricultural and Forest Meteorology* 182–183, 67–75.

TeKrony, D.M. and Egli, D.B. (1997). Accumulation of seed vigour during development and maturation. In: Ellis, R.H., Black, M., Murdock, A.J. and Hong, T.D. (eds) *Basic and Applied Aspects of Seed Biology*. Kluwer Academic Publishers, London, UK, pp. 369–384.

TeKrony D.M. and Hunter J.L. (1995). Effect of seed maturation and genotype on seed vigor in maize. *Crop Science* 35, 857–862.

TeKrony, D.M., Egli, D.B., Balles, J., Pfeiffer, T. and Fellows, R.J. (1979). Physiological maturity in soybean. *Agronomy Journal* 71, 771–775.

TeKrony, D.M., Egli, D.B. and Henson, G. (1981). A visual indicator of physiological maturity in soybean plants. *Agronomy Journal* 73, 553–556.

Thies, J.E., Singleton, P.W. and Bohtool, B.B. (1995). Phenology, growth and yield of field-grown soybean and bush bean as a function of varying modes of N nutrition. *Soil Biology and Biochemistry* 27, 573–583.

Thomaier, S., Specht, K., Heuckel, D., Dierich, A., Siebert, R. and Freisinger U.B. (2015). Farming in and on urban buildings: Present practice and specific novelties of zero-acreage farming (ZFarming). *Renewable Agriculture and Food Systems* 30, 43–54.

Thomas, H. and Stoddart, J.L. (1980). Leaf senescence. *Annual Review of Plant Physiology* 31, 83–111.

Thomison, P.R. and Jordan, D.M. (1995). Plant population effects on corn hybrids differing in ear growth habit and prolificacy. *Journal of Production Agriculture* 8, 394–400.

Thompson, J.F., Madison, J.T. and Muenster, A.E. (1977). *In vitro* culture of immature cotyledons of soya bean (*Glycine max* L. Merrill). *Annals of Botany* 41, 29–39.

Thorne, J.H. (1981). Morphology and ultrastructure of maternal seed tissues of soybean in relation to the import of photosynthate. *Plant Physiology* 67, 1016–1025.

Thorne, J.H. (1985). Phloem unloading of C and N in developing seeds. *Annual Review of Plant Physiology* 36, 317–343.

Thornley, J.H.M. (1980). Research strategy in the plant sciences. *Plant, Cell and Environment* 3, 233–236.

Timsina, J. and Connor, D.J. (2001). Productivity and management of rice–wheat cropping systems: Issues and challenges. *Field Crops Research* 69, 93–132.

Tollenaar, M. (1989). Genetic improvement in grain yield of commercial maize hybrids grown in Ontario from 1959 to 1988. *Crop Science* 29, 1365–1371.

Tollenaar, M. and Bruulsema, T.W. (1988). Effects of temperature on rate and duration of kernel dry matter accumulation of maize. *Canadian Journal of Plant Science* 68, 935–940.

Tollenaar, M. and Daynard, T.B. (1978). Kernel growth and development of two positions in the ear of maize (*Zea mays*). *Canadian Journal of Plant Science* 58, 189–197.

Tollenaar, M. and Wu, J. (1999). Yield improvement in temperate maize is attributable to greater stress tolerance. *Crop Science* 39, 1597–1604.

Tollenaar, M., Dwyer, L.M., Stewart, D., W, and Ma, B.L. (2000). Physiological parameters associated with differences in kernel set among maize hybrids. In: Westgate, M.E. and Boote, K.J. (eds) *Physiology and Modeling Kernel Set in Maize*. ASA-CSSA, Madison, Wisconsin, USA, pp. 115–130.

Trewatha, S.E., TeKrony, D.M. and Hildebrand, D. (1993). Lipoxygenase activity and C_6-aldehyde formation in comparison to germination and vigor during soybean seed development. *Crop Science* 33, 1337–1344.

Uhart, S.A. and Andrade, F.A. (1995a). Nitrogen deficiency in maize. II. Carbon–nitrogen interaction effects on kernel number and grain yield. *Crop Science* 35, 1384–1389.

Uhart, S.A. and Andrade, F.H. (1995b). Nitrogen and carbon accumulation and remobilization during grain filling in maize under different source/sink ratios. *Crop Science* 35, 183–190.

Unger, P.W. and Thompson, T.E. (1982). Planting date effects on sunflower head and seed development. *Agronomy Journal* 74, 389–395.

United Nations (2015). *World Population Prospects: The 2015 Revision: Key Findings and Advance Tables*. United Nations Department of Economic and Social Affairs, Working Paper No. ESA/P/WP241. Available at http://esa.un.org/unpd/wpp/Publications/Files/Key_Findings_wpp_2015.pdf (accessed 27 August 2016).

Van Camp, W. (2005). Yield enhancement genes: Seeds for growth. *Current Opinions in Biotechnology* 16, 147–153.

Van de Venter, H.A., Demir, I., de Meillon, S. and Loubser, W.A. (1996). Seed development and maturation in edible dry bean (*Phaseolus vulgaris* L.). *South African Journal of Plant and Soil* 13, 47–50.

Vanham, D., Bouraoui, F., Leip, A., Grizzette, B. and Bidoglio, G. (2015). Lost water and nitrogen resources due to EU consumer food waste. *Environmental Research Letters* 10. DOI: 10.1088/1748–9326/10/8/084008.

Van Sanford, D.A. (1985). Variation in kernel growth characters among soft red winter wheats. *Crop Science* 25, 626–630.

van Schaik, P.H. and Probst, A.H. (1958). The inheritance of inflorescence type, peduncle length, flowers per node and percent flower shedding in soybeans. *Agronomy Journal* 50, 98–102.

Vaughan, J.G. and Geissler, C.A. (1997). *The New Oxford Book of Food Plants*. Oxford University Press, Oxford, UK.

Vega, C.R.C., Andrade, F.H., Sadras, V.O., Uhart, S.A. and Valentinuz, O.R. (2001). Seed number as a function of growth: A comparative study in soybean, sunflower, and maize. *Crop Science* 41, 748–754.

Venkateswarlu, B., Rao, J.S. and Rao, A.V. (1977). Relationship between growth duration and yield parameters in irrigated rice (*Oryza sativa*). *Indian Journal of Plant Physiology* 20, 69–76.

Vieira, R.D., TeKrony, D.M. and Egli, D.B. (1992). Effect of drought and defoliation stress in the field on soybean seed germination and vigor. *Crop Science* 32, 471–475.

Villalobos, F.J., Sadras, V.O., Soriano, A. and Fereres, E. (1994). Planting density effects on dry matter partitioning and productivity of sunflower hybrids. *Field Crops Research* 36, 1–11.

von Caemmerer, S., Quick, W.P. and Furbank, R.T. (2012). The development of C4 rice: current progress and future challenges. *Science* 336, 1671–1672.

Vos, J. (1981). Effects of temperature and nitrogen supply on post-floral growth of wheat: Measurements and simulations. Centre for Agricultural Publishing and Documentation, Wageningen, The Netherlands.

Wada, G. and Cruz, P.L.S. (1989). Varietal differences in nitrogen response of rice plants with special reference to growth duration. *Japanese Journal of Crop Science* 58, 732–739.

Walbot, V. (1978). Control mechanisms for plant embryogeny. In: Clutter M.E. (ed.) *Dormancy and Developmental Arrest.* Academic Press, London, UK, pp. 113–166.

Walker, G.K., Miller, M.H. and Tollenaar, M. (1988). Source–sink limitations of maize growing in an outdoor hydroponic system. *Canadian Journal of Plant Science* 68, 947–955.

Wallace, S.U. (1986). Yield and seed growth at various canopy locations in a determinate soybean cultivar. *Agronomy Journal* 78, 173–178.

Wallace, J.S. (2000). Increasing agricultural water use efficiency to meet future food production. *Agricultural Ecosystems & Environment* 82, 105–119.

Wardlaw, I.F. (1970). The early stages of grain development in wheat: Response to light and temperature in a single variety. *Australian Journal of Biological Science* 23, 765–774.

Wardlaw, I.F. (1990). The control of carbon partitioning in plants. *New Phytologist* 116, 341–381.

Wardlaw, I.F. and Moncur, L. (1995). The response of wheat to high temperatures following anthesis. I. The rate and duration of kernel filling. *Australian Journal of Plant Physiology* 22, 391–397.

Wardlaw, I.F., Sofield, I. and Cartwright, P.M. (1980). Factors limiting the rate of dry matter accumulation in the grain of wheat grown at high temperatures. *Australian Journal of Plant Physiology* 7, 387–400.

Warrag, M.O.A. and Hall, A.E. (1984). Reproductive responses of cowpea (*Vigna unguiculata* (L.) Walp.) to heat stresses. II. Response to night air temperature. *Field Crops Research* 8, 17–33.

Warren, J.M. (2015). *The Nature of Crops: How We Came to Eat the Plants We Do.* CAB International, Wallingford, Oxfordshire, UK.

Watson, D.J. (1947). Comparative physiological studies on the growth of field crops. I. Variation in net assimilation rate and leaf area between species and varieties, and within and between years. *Annals of Botany* 11, 41–76.

Watson, D.J. (1958). The dependence of net assimilation rate on leaf-area index. *Annals of Botany* 22, 37–54.

Watson, P.A. and Duffus, C.M. (1991). Light-dependent CO_2 retrieval in immature barley caryopsis. *Journal of Experimental Botany* 42, 1013–1020.

Weber, H., Heim, U., Golombek, S., Borisjuk, L. and Wobus, U. (1998). Assimilate uptake and regulation of seed development. *Seed Science Research* 8, 331–346.

Wells, R., Schulze, L.L., Ashley, D.A., Boerma, H.R. and Brown, R.H. (1982). Cultivar differences in canopy apparent photosynthesis and their relationship to seed yield in soybean. *Crop Science* 22, 886–890.

Wennblom, R.D. (1978). Have crop yields peaked out? *Farm Journal* 102, 32, 34.

West, T.D., Griffith, D.R., Steinhardt, G.C., Kladirko, E.J. and Parsons, S.D. (1996). Effect of tillage and rotation on agronomic performance of corn and soybean: Twenty year study on dark silty clay loam soil. *Journal of Production Agriculture* 9, 241–248.

Westgate, M.E. (1994). Water status and development of the maize endosperm and embryo during drought. *Crop Science* 34, 76–83.

Westgate, M.E. and Grant, D.T. (1989a). Effects of water deficits on seed development in soybean. I. Tissue water status. *Plant Physiology* 91, 975–979.

Westgate, M.E. and Grant, D.T. (1989b). Water deficit and reproductive development in maize. Response of the reproductive tissues to water deficits at anthesis and mid-grain fill. *Plant Physiology* 91, 862–867.

Westgate, M.E., Schussler, J.R., Reicosky, D.C. and Brenner, M.L. (1989). Effects of water deficits on seed development in soybean. II. Conservation of seed growth rate. *Plant Physiology* 91, 980–985.

Whan, B.R., Carlton, G.P. and Anderson, W.K. (1996). Potential for increasing rate of grain growth in spring wheat. *Australian Journal of Agriculture Research* 47, 17–31.

Wheeler, T.R., Hong, T.D., Ellis, R.H., Batts, G.R., Morrison, J.I.L., and Hadley, P. (1996). the duration and rate of grain growth and harvest index of wheat (*Triticum aestivum*) in response to temperature and CO_2. *Journal of Experimental Botany* 47, 623–630.

White, J. (1981). The allometric interpretation of the self-thinning rule. *Journal of Theoretical Biology* 89, 475–500.

White, J.W. and Izquierdo, J. (1991). Physiology of yield potential and stress tolerance. In: van Schoonhover, A. and Voysest, O. (eds) *Common Beans: Research for Crop Improvement.* CAB International, Wallingford, Oxfordshire, UK, pp. 287–382.

Whitfield, D.M. (1992). Effects of temperature and ageing on CO_2 exchange of pods of oil seed rape (*Brassica napus*). *Field Crops Research* 28, 271–280.

Whitfield, D.M., Conner, D.J. and Hall, A.J. (1989). Carbon dioxide balance of sunflower (*Helianthus annus* L.) subjected to water stress during grain filling. *Field Crops Research* 20, 65–80.

Wien, H.C. and Ackah, E.E. (1978). Pod development period in cowpeas: Varietal differences as related to seed characteristics and environmental effects. *Crop Science* 18, 791–794.

Wilhelm, E.P., Mullin, R.E., Keeling, P.L. and Singletary, G.W. (1999). Heat stress during grain filling in maize: Effects on kernel growth and metabolism. *Crop Science* 39, 1733–1741.

Wilson, J.W. (1967). Ecological data on dry matter production by plants and plant communities. In: Bradley, E.F. and Denmead, O.T. (eds) *The Collection and Processing of Field Data.* Interscience Publishers, New York, USA, pp. 77–123.

Wilson, L.A. (1977). Root crops. In: de T. Alvin, P. and Kozlowski, T.T. (eds) *Ecophysiology of Tropical Crops.* Academic Press, New York, USA, pp. 187–236.

Wilson, R.F. (1987). Seed metabolism. In: Wilcox, J.R. (ed.) *Soybeans: Improvement, Production, and Uses* 2nd edn. ASA-CSSA-SSSA, Madison, Wisconsin, USA, pp. 643–686.

Wilson, R. (2004). Seed composition In: Boerma, H.R. and Specht, J.E. (eds) *Soybeans: Improvement, Production, and Uses,* 3rd edn. ASA-CSSA-SSSA, Madison, Wisconsin, USA, pp. 621–677.

Wittenbach, V.A. (1983). Effect of pod removal on leaf photosynthesis and soluble protein composition of field grown soybeans. *Plant Physiology* 73, 123–124.

Wolf, D.W., Henderson, D.W., Hsiao, T.C. and Alvino, A. (1988a). Interactive water and nitrogen effects on senescence of maize. I. Leaf area duration, nitrogen distribution and yield. *Agronomy Journal* 80, 859–864.

Wolf, D.W., Henderson, D.W., Hsiao, T.C. and Alvino, A. (1988b). Interactive water and nitrogen effects on senescence of maize. II. Photosynthesis decline and longevity of individual leaves. *Agronomy Journal* 80, 865–870.

Wolswinkel, P. and deRuiter, H. (1985). Amino acid release from the seed coat of developing seeds of *Vicia faba* and *Pisum sativum. Annals of Botany* 55, 283.

Wych, R.D., McGraw, R.L. and Suthman, D.D. (1982). Genotype × year interaction for length and rate of grain filling in oats. *Crop Science* 22, 1025–1028.

Wyss, C.S., Czyzewicz, J.R. and Below, F.E. (1991). Source–sink control of grain composition in maize strains divergently selected for protein concentration. *Crop Science* 31, 761–766.

Yang, J., Zhang, L., Huang, Z., Zhu, Q. and Wang, L. (2000). Remobilization of carbon reserves is improved by a controlled soil-drying during grain filling of wheat. *Crop Science* 40, 1645–1655.

Yang, J., Zhang, J., Wang, Z. and Zhu, Q. (2001). Activities of starch hydrolytic enzymes and sucrose–phosphate synthase in the stems of rice subjected to water stress during grain filling. *Journal of Experimental Botany* 52, 2169–2179.

Yang, L., Liu, H., Wang, Y., Zhu, J., Huang, J., Liu, G., Dong, G. and Wang, Y. (2009). Impact of elevated CO_2 concentration on inter-subspecific hybrid rice cultivar Liangyoupeijiu under fully open-air field conditions. *Field Crops Research* 112, 7–15.

Yang, X., Chen, F., Lin, X., Lin, Z., Zhang, H., Zhao, J., Li, K., Ye, Q., Li, Y., Lv, S. *et al.* (2015). Potential benefits of climate change for crop productivity in China. *Agricultural and Forest Meteorology* 208, 76–84.

Yang, C., Sui, R. and Lee, W.S. (2016). Precision agriculture in large-scale mechanized farming. In: Zhnag, Q. (ed.) *Precision Agriculture for Crop Farming.* CRC Press. Boca Raton, Florida, USA, pp. 117–212.

Yazdi-Samadi, B., Rinne, R.W. and Seif, R.D. (1977). Components of developing soybean seeds: Oil, protein, sugars, starch, organic acids, and amino acids. *Agronomy Journal* 69, 481–486.

Yoshida, S. (1977). Rice. In: Alvim, P.D.T. and Kozlowski, T.T. (eds) *Ecophysiology of Tropical Crops.* Academic Press, New York, USA, pp. 57–87.

Yoshida, S. and Hara, T. (1977). Effects of temperature and light on grain filling of an indica and japonica rice (*Oryza sativa* L.) under controlled environmental conditions. *Soil Science and Plant Nutrition* 23, 93–107.

Yoshida, K., Nomma, F. and Gotoh, K. (1983). Significance of intra-plant flowering date on soybean seed production. I. Pod and seed development among different flowering dates. *Japanese Journal of Crop Science* 52, 555–561.

Yu, Z.-W., van Sanford, D.A. and Egli, D.B. (1988). The effect of population density on floret initiation, development and abortion in winter wheat. *Annals of Botany* 62, 295–302.

Zahedi, M. and Jenner, C.F. (2003). Analysis of the effects of high temperature on the grain filling attributes of wheat estimated from mathematical models of grain filling. *Journal of Agricultural Science* 141, 203–212.

Zeiher, C., Egli, D.B., Leggett, J.E. and Reicosky, D.A. (1982). Cultivar differences in N redistribution in soybeans. *Agronomy Journal* 74, 375–379.

Zhang, O. (2007). Strategies for developing green super rice. *Proceedings of the National Academy of Sciences* 104, 16402–16409.

Zhang, H., Tan, G., Yang, L., Yang, J., Zhang, J. and Zhao, B. (2009). Hormones in the grains and roots in relation to post-anthesis development of inferior and superior spikelets in japonica/indica hybrid rice. *Plant Physiology and Biochemistry* 47, 195–204.

Zinselmeier, C., Westgate, M.E., Schussler, J.R. and Jones, R.L. (1995). Low water potential disrupts carbohydrate metabolism in maize (*Zea mays* L.) ovaries. *Plant Physiology* 107, 385–391.

Index

Note: Page numbers in **bold** type refer to **figures**
Page numbers in *italic* type refer to *tables*